ILLUSTRATED GUIDE TO JEWELRY APPRAISING

Illustrated Guide to Jewelry Appraising

ANTIQUE, PERIOD, AND MODERN

ANNA M. MILLER, GG, ASA

CHAPMAN & HALL
New York • London

ISBN 0-412-98931-X

This edition published by
Chapman & Hall
One Penn Plaza
New York, NY 10119

Published in Great Britain by
Chapman & Hall
2-6 Boundary Row
London SE1 8HN

Printed in the United States of America

16 15 14 13 12 11 10 9 8 7 6 5 4 3 2

Library of Congress Cataloging in Publication Data

Miller, Anna M., 1933-.
Illustrated guide to jewelry appraising : antique, period, and modern / Anna M. Miller.
p. cm.
Bibliography: p.
Includes index.
ISBN 0-442-31944-4
1. Jewelry—Valuation. I. Title.
NK7304.M5 1990
739.27'029'7—dc20 89-16452
CIP

British Library Cataloguing in Publication Data available

Please send your order for this or any **Chapman & Hall book to Chapman & Hall, 29 West 35th Street, New York, NY 10001, Attn: Customer Service Department.** You may also call our Order Department at 1-212-244-3336 or fax your purchase order to 1-800-248-4724.

For a complete listing of Chapman & Hall's titles, send your requests to **Chapman & Hall, Dept. BC, One Penn Plaza, New York, NY 10119.**

To Joe, for his help
and encouragement

CONTENTS

PREFACE

There is more to appraising jewelry than just being able to put a dollar value on an item. The title of appraiser distinguishes the individual who is able to identify, witness, estimate status, excellence, or potentiality, and to determine the authenticity of an article. Many factors impact on a thorough appraisal, especially on estate and period jewelry.

Developing all this expertise is a lot to ask of practitioners who only a decade ago were barely making a distinction between a well-written sales receipt and a professional appraisal report.

In the past few years bold changes have taken place in this field. It is now understood that standard appraisal concepts and principles can be applied to the appraisal of jewelry in the same way they are applied to real and other personal property.

One thing has been learned for certain. The jewelry appraiser needs historical knowledge of jewelry along with information on modern trends. Jewelry is a great mirror of a society and the people of a particular culture, and reflects the taste and attitudes of a period. It is interesting to note that the value of an item varies from the relationship of the item to its owner and his culture, customs, and geographical location. In our present society, it becomes clear that an item is also measured against the social, economic, religious, and environmental forces that affect value.

We must also consider that, while technical equipment is necessary for proper identification of gemstones and metals in most cases, more than those tests alone contribute to estimating market value of an object. A test of appraiser competence is demonstrated by how well he or she is able to *read the jewelry*. How well the appraiser is able to define the manufacturing process, quality grade the gemstones, assess the design for balance, harmony, and rhythm, circa date, and analyze the local, regional, and national market for comparable sales and a value determination. Then, all this information, with clear jewelry descriptions, must be succinctly put together with photographs and delivered to the client.

Today's jewelry appraiser should also be cognizant of the vicissitudes of fashion, how changes impact the jewelry market in a substantial and vital manner in both design and style. A careful look must be given to study of color psychology with an awareness of why specific gemstones and their colors and enamels of certain colors were used in different periods. Industrial developments, from machine stamped jewelry, the use of aniline dyes in clothing, and development of the electric light, have influenced design and use of gemstones. Jewelry has moved from an expression of wealth and power to a way of communicating personal feelings. Having this knowledge not only adds to the expertise needed to date jewelry, but also helps to authenticate it. Willing or not, the appraiser is also jewelry historian. In an increasingly mobile and disposable culture, in a period of time where a borderless European continent may become a reality, more people than ever want a direct link to the past and wish to know about their ancestors and roots. This is evident by the growing interest and sales of antique and period jewelry. A sudden plethora of dealers in antique and estate jewelry, magazines, books, and seminars dedicated to period jewelry have fanned the flame of consumer interest.

Finding a dollar value for any item of jewelry is not a simple task. At the present time, there is no Blue Book of jewelry prices. Calculating a sum requires a complicated series of steps and inquiries with answers based on the appraiser's ability to correlate, analyze, and do deductive reasoning. Although old fashioned legwork still plays a stellar role, appraisers cannot just run amuck in the market and must know what information to collect, where to find it, and how to interpret the data. Helping appraisers frame the question of what information is necessary, and then helping find the answer is one of the aims of this book.

Other purposes are to reiterate and improve on basic principles of jewelry appraising, to refine and expand the nomenclature, to guide both novice and practicing appraiser to a better understanding of antique and period jewelry, to suggest clues for more accurate circa dating, to assist in writing concise narrative descriptions, to point out a practical methodology for research, and to assemble information needed for an accurate estimation of value.

The number of jewelry appraisers is increasing, which results in a collective excitement and broader expertise in this field. The development of the discipline is far from over. As it grows, new practitioners will bring their ideas and insights into this arena. We applaud their efforts.

It is hoped this book will provide basic information to build upon and will be of genuine assistance in the continued development and practice of jewelry appraisal, today and into the future.

Prices in this book should be viewed only as general guides, as they are conditional to the date of this writing and valid in the author's regional market. Prices may be higher or lower in other geographic localities.

ACKNOWLEDGMENTS

Many colleagues and friends contributed time and information to this book. First, thanks to Joanna Angel for her inspiring enthusiasm and knowledge.

Grateful thanks to those who supplied invaluable information: Patrice Phillips of the Diamond Information Center, Reginald Miller, Lazar Kaufman, Moshe Kaufman, Albert Asher, Robert Sandler, Mark Sandler, Steven Orgel, Howard Rubin, Alison LeBaron, Susan Eisen, Rupa Dutia, Consuelo White, Mark P. Moeller, Charlene Fischman, John Miller, Karen Lorene, Lloyd Lieberman, Carolyn Price Farouki, Marcia Mayo, Therese Kienstra, Sue Johnson, GIA; Dona M. Dirlam, GIA; George Houston, Rebecca Baker of House of Onyx; Peggy Blackford, Ed Menk, Mark Valente, Cortney Balzan, Doug Sparrow, Alan Revere, James V. Jolliff, Pat deRobertis, Edith Weber, Myra Waller, Richard Raymond Alasko, Pam Abramson, Tony Valente, Joyce Jonas, and Neil Cohen.

Special thanks to Tom R. Paradise for the many helpful facts from his special fund of knowledge.

I particularly want to thank Sara Levy for allowing unlimited access to her jewelry for photographing and to Ellen Epstein for permitting me to take a family heirloom cross-country to be photographed.

I wish to extend thanks to many members of the American Society of Appraisers who by their confidence in this work made completion of the book possible, with special mention to the Personal Property and Gems and Jewelry disciplines.

Thanks to members of the Southern Women's Jewelry Association and the Association of Women Gemologists for their staunch moral support and encouragement. To Elizabeth Hutchinson for the illustrations and to Van Edwards for both the cover photograph and all other photographs not otherwise credited, thank you.

To my Japanese colleagues, Toshio Ishida and Tadaaki Saito, your cooperation and willingness to provide information was much appreciated.

Finally, thanks to Joseph W. Tenhagen, ASA, MGA; and to Dr. Charles D. Peavy, ASA; for their advice and counsel and steadfast belief in the importance of this work.

Anna M. Miller, GG, ASA
Master Gemologist Appraiser

CHAPTER 1

THE PROFESSIONAL APPRAISER

"Knowledge is of two kinds.
We know a subject ourselves,
or we know where we can find
information upon it."
Samuel Johnson

In appraising, knowledge is power only if it can be communicated articulately. If the three-word mantra of the real estate appraiser is location, location, location, then the chant of the jewelry appraiser must be research, research, research, with education, experience, and connoisseurship added as support and foundation.

Historically, jewelry appraising as a profession is in its infancy, but the same basic concepts and principles for responsible jewelry appraising have been around for a long time. When law divided material properties into "real" and "personal," real property assumed a dominant role in business matters. Personal property, called chattels, was of lesser importance. Thus, real property (real estate) appraisers have dominated valuation practices despite the vast financial investments in personal property and the valuation problems inherent in buying, selling, and insuring these goods.

As a growing awareness of investment potential in arts, antiques, machinery, tools, stamps, coins, sculpture, glass, china, household furnishings, gems, and jewelry emerged, personal property appraisers began to read, study, modify, and codify existing defined standards. They realized that the differences between the appraisal specialties that made them appear to be quite unrelated were in the kinds of *data* and property attributes with which the appraisal was concerned and *not* in the principles and methods. Henry A. Babcock wrote in "Appraisal, Principles, and Procedures" that the principles and methods of valuation for a diamond ring and a single-family residence were one and the same—the differences were between the technological data. Appraisers of diamonds must know how diamond purchasers react to the four Cs of cut, clarity, color, and carat weight, while the residence appraiser is concerned with buyer reaction to the value elements of square footage, number of rooms, age of the dwelling, and location.

Until recently the jewelry appraiser has avoided using any standard appraisal structure and has voiced indifference to using the existing methodologies, concepts, and principles. The awakening has begun, and today many gems and jewelry appraisers are clamoring for precise valuation information. There lies ahead a great development period in which the jewelry appraiser will recognize the tie between appraisal principles and their unifying concepts, which pervade and are basic to all branches of the profession.

The gems and jewelry appraiser is in a period of evolution and maturation; he or she must understand that, even with the knowledge of special techniques, the business is complex and volatile. The realization that estimates of value assigned to items will materially impact both client and subject for a long time is beginning to get the appraiser's attention and consideration. Thus, more and more valuers are concluding that appraising must be more than just a sideline of the jeweler or gemologist. It demands full status as a profession.

The subject of valuation is an alive and vital topic that requires the student and practitioner of valuation science to keep current with trends by a continuing education process. The lament most often heard even from conscientious appraisers is: "I thought I was at the top of my profession, then someone moved the top!" Just keeping up with changes and trends in one's own appraisal specialty is not enough. In this fast-paced society, the jewelry appraiser needs to have scholarship in fine arts, decorative arts, antiques, fashion, coins, and stamps, among others. The deeper one delves into valuation science, the clearer it becomes that a common thread connects all personal property. In Chapter Three, attention is drawn to ways in which a knowledge of social customs and cultures, costumes, motifs, and designs of times past can aid today's jewelry appraiser with identification and value-determination information.

Though it may sound farfetched, the study and knowledge of worldwide social, economic, political, and environmental conditions are becoming essential for the appraiser. These forces have *direct impact* on the value of gemstones and jewelry. The great scope of the appraisal field embraces many ideas and concepts.

Being able to interpret a fluctuating marketplace is unquestionably vital. Last year's appraisal values may be drastically different when one considers this year's politics, environmental changes, and inflation or recession economics. Those who doubt the impact of foreign policy upon their local gemstone values have only to reflect on how the political unrest in southeast Asia closed ruby and sapphire mines and caused those gems to skyrocket in price, or to study the price impact on precious gemstones and precious metals when there is news of mine depletions. Inflation often creates a scenario in which poorer quality gems and jewelry are offered instead of price increases. The new synthetic gemstones are another problem area. The competent appraiser needs training in the use of modern equipment to be able to distinguish the synthetics constantly arriving in the market. Indeed, just being able to identify heirloom jewelry from fine reproductions speaks well of the competence of the appraiser!

Appraisers need access to actual pricing information at all levels of the marketplace. Currently, they continue to establish value from researched sources in their own geographical area, while pricing information at an international level is becoming more essential, especially for gem-quality goods. One reason appraisers need to be aware of international value is simple: Fine gemstones and well-crafted jewelry prices vary less geographically as their market values increase and the circle of buyers and sellers shrinks. At the highest market levels, the prices paid and the values assigned will often be one and the same regardless of geographical location. This can be accounted for because of the excellent communications in the jewelry industry, augmented by the use of computers, FAX communication, and telecommunication networks. Designer jewelry needs special attention in price research and analysis.

Primary Function of the Appraiser

The starting point of any appraisal is an investigation of the property to be appraised. The appraiser has a primary function when he or she contracts to make a valuation, and that is to determine what level of the market is most appropriate for the purpose of the item being appraised. The appraiser must be aware of any constraints of that market (such as scarcity). There are several kinds of value and each has its legitimate place for some type of jewelry appraisal. Retail replacement value (new and used), fair market value, and liquidation value are the most often sought. Of course, the appraiser will take the client's instructions and/or wishes into consideration, but in the final analysis it is the appraiser's sole responsibility to select the appropriate value or estimated cost. It is also the responsibility of the appraiser to convey and fully explain to the client what is meant by the particular value that has been determined in order to prevent unwitting or deliberate misapplication of the appraisal.

For instance, an appraisal that calls for replacement value of an antique brooch for insurance would not be fulfilled properly by an appraisal of retail replacement value as a new piece of jewelry. Vintage jewelry can only be compared with other (used) comparable or identical vintage jewelry. A summary of related costs and values is illustrated in table 1-1, Purpose and Function of Appraisals. The appraiser must keep in mind that he or she must perform with a clear awareness of the public confidence that rests on his or her integrity and knowledge of valuation techniques. In the final analysis, the most competent appraiser may be the person who knows his or her limitations. The functions of an appraiser can be outlined in this manner:

1. To determine monetary value of a particular item of jewelry or gemstone
2. To determine the quality of the property
3. To establish the appropriate market for the subject being valued
4. To identify and authenticate if possible and if required
5. To estimate the cost of replicating the property, if it is the only means of replacement
6. To reach conclusions and recommendations

Primary Objective of the Appraiser

It is the client who has a purpose in having an appraisal made; the *appraiser* defines the objective. A client's desire may be to obtain a collateral loan on jewelry; thus the appraiser's objective is to determine the current fair market value of the items. The appraiser's objective is the *end point,* the final estimate of value, which is determined by research and by following standard procedure to its conclusion; that is, to analyze and estimate the market value of an item at a specific time and place for a particular purpose. There are three classes of appraisal objectives:

- Qualitative
- Quantitative (nonmonetary)
- Monetary

Qualitative appraisal has as its end point a conclusion or opinion of the quality of an item, such as the quality of a diamond stated in terms of cut, color, clarity, and carat weight. The appraiser would also express an opinion as to the make of the stone and the quality of the craftsmanship on any attendant mounting. This class of appraisal is expressed in an item's condition, quality, authenticity, and opinion of artistic merit.

Quantitative nonmonetary appraisal may use figures in a nonmonetary way, such as expressing an opinion on the age of an item or circa dating (approximate dating of an item plus or minus ten years). This is estimated data and most often is expressed in an appraisal in terms of age, amount of natural resources, or supply and demand.

Monetary appraisals deal with estimates of value and are the type most often seen by jewelry appraisers. The primary objective in a monetary appraisal is to determine a numerical result or most probable point value.

Let's briefly ponder exactly why we do appraisals. Basically, the narrative report of monetary value on a jewelry item has six functions:

1. To identify the subject
2. To specify the market in which the item is being valued
3. To name and support the choice of value approach
4. To research and establish supporting documentation for the numerical results
5. To offer full narrative description of the article in such a manner that it can be easily understood and reproduced in identical kind if necessary
6. To determine and be able to support provenance

Richard Rickert, professor of Valuation Science at Lindenwood College, St. Charles, Mo., says if the identification of the item is not accurately and sufficiently descriptive, no one can tell which features of the jewelry generate its value or how they can be distinguished from all similar properties. In addition, a complete identification and description of the property helps the client or other reader of the report to make a better judgment on the validity of the appraiser's valuation. Thorough descriptions also help to prevent the deceitful use of the appraisal to represent a similar piece of jewelry that may be less valuable than the original item appraised. Descriptions that are sparse and inaccurate lead to uncertain conclusions and indefensible valuations.

Value Definitions and Their Application to Jewelry Appraisals

Value in gems and jewelry varies depending upon the relationship of the item to the person, to the culture, and to the social, regional, and geographical location. Most appraisers will content themselves with research on the value of an item in their region, and concentrate on establishing a *mode* (most often occurring price in a series of prices) that can be tracked and documented. There is a great deal of confusion over value, price, and cost. Appraisers try to attach a measurement of some kind to items so that a value can be established. *Price* and *cost* refer to an amount of money asked or actually paid for a piece of jewelry, and this may be more or less than its *value.* The relationship between price and value is often misunderstood.

Production cost of an item of jewelry may not be its intrinsic *value;* rather it is the estimated amount of the cost to produce a piece of jewelry by assembling the separate components, manufacturing it, and finishing it. Also, a buyer's purchase price of an item may not be the true value. Price, commonly referred to as a sale, is generally an accomplished fact, while value is usually an *expected monetary* amount that should result under specific market conditions. Without an appraisal of an item, it is not known whether a price actually paid or received equaled the jewelry's market value. Even though actual prices can provide evidence of market value, the appraiser must analyze sales transaction prices before reaching a value conclusion.

The entire fabric of appraisal principles and procedures is woven from definitions of property, costs, and concepts of the value of property. There are several kinds of value:

Market Value, an economic concept, is the amount of money a willing seller can expect to obtain for an item from a willing buyer. Market value has been defined as "the most probable price in cash, or in other precisely revealed terms, for which the appraised property will sell in a competitive market under all conditions requisite to a fair sale, with the buyer and seller each acting prudently, knowledgeably, and for self-interest, and assuming that neither is under undue duress." Implicit in this statement is the completion of a sale at a specific date and the passing of an item from seller to buyer under the following conditions:

1. The buyer and seller are equally motivated.
2. Both parties are well informed and/or advised and each is acting in what he or she considers his or her best interest.
3. A reasonable time has been allowed for exposure of the item to the open market.
4. Payment is made in cash or its equivalent.
5. Financing, if any, is available and typical for the locale.
6. The price represents normal sale for the item and is not affected by any special fees or costs.
7. The sale and transfer of goods are normal and predictive for the type of jewelry.

Retail Value as it applies to jewelry appraisals is usually considered to mean retail replacement value for insurance replacement. This is the price at which the average retail jeweler would sell a new item that is comparable to the item being appraised at the time of appraisal. Also, it may be defined as retail replacement (used) and as the average price for antique, estate, second-hand, and used jewelry. It may also be the price (with restrictions) for certain types of jewelry pieces by well-known designers and/or jewelry manufacturers and stamped with their signature or trademark.

Wholesale Value is the price a retail jeweler expects to pay from a jewelry manufacturer or gemstone importer for items he expects to resell.

Replacement Value is the cost of replacing or duplicating *exactly* an item of jewelry at the current cost and time in exact quality and/or utility.

Auction Value is the price for jewelry sold to the highest bidder at an auction. Actual realized prices may be used to determine appraisal values when an item is being valued for replacement or liquidation purposes. With *careful* qualifying statements and substantiating documentation, auction prices can be used in fair market value appraisals.

Liquidation is the cash price that may be expected when selling a jewelry item *usually* under duress and/or at a forced sale.

Fair Market Value is a legal term; the Treasury Regulation and definition of fair market value (FMV) is:

> The fair market value is the price at which the property would change hands between a willing buyer and a willing seller, neither being under any compulsion to buy or sell and both having reasonable knowledge of relevant facts. The fair market value of a particular item of property includible in the decedent's gross estate is not to be determined by forced sale price. Nor is the fair market value of an item of property to be determined by the sale price of the item in a market other than that in which such item is most commonly sold to the public, taking into account the location of the item wherever appropriate. Thus, in the case of an item of property includible in the decedent's gross estate, which is generally obtained by the public in the retail market, the fair market value of such an item of property is the price at which the item or a comparable item would be sold at retail. TR 20.2031-1(b)

A comprehensive discussion and interpretation of this market can be found in the author's book *Gems and Jewelry Appraising: Techniques of Professional Practice.*

Scrap Value is the lowest measurement of monetary return and may also be referred to as melt value. It is the monetary return on the materials (gold, silver, platinum) of the item without considering any other purposes or aesthetic criteria.

Purpose and Function of an Appraisal

The first step in writing an appraisal report is to determine the reason for the document. The purpose and function alert the appraiser to the correct valuation approach. Table 1-1 diagrams Purpose and Function and its relation to various values. Explanations of the terms are included.

Table 1-1. Purpose and Function of Appraisals

Purpose	Function
Retail Replacement can be defined on two levels: Retail replacement new means the cost in the current and prevailing regional marketplace to replace an appraised item of jewelry with an identical new article of exact utility and quality. Replacement cost with a comparabie article means the cost in the current and prevailing regional marketplace that allows an item to be replaced with an exact replica to that being appraised, or comparable item equal in quality and utility to that appraised.	**Insurance** appraisal is required by insurance companies before they will insure jewelry beyond certain dollar amounts. The value is determined with the current researched market price of a comparable item in the appraiser's locale. The value is derived from research of numerous comparable sales, not the policies or prices of one retail jeweler. **Comparison** appraisal is frequently a quality report listing the essential quality features of an item of jewelry. This report/appraisal may or may not have monetary values included, depending upon the needs of the client. Comparison appraisals are often asked for to justify a sales price. **Hypothetical** appraisal is made on an item of jewelry that may not have been seen by the appraiser. The item may no longer exist (theft, fire, disappearance). These appraisals are based on information supplied by the client. There may be partial proof, such as a previous appraisal or bills of sale. Photographs, if available, will help solidify the basis of the work. **Damage** appraisal determines how damage occurred to an item, the extent of the damage, and the possible repair alternatives. **Barter** appraisals must narrate various market levels and the possibility of sale of the appraised goods on those markets. This is a sensitive market to address, and the appraiser must also investigate the highest and best use potential of the goods for the client to be fully informed.
Fair Market Value is defined by the Treasury Regulations as "The price at which the property would change hands between a willing buyer and a willing seller, neither being under any compulsion to buy or sell and both having a reasonable knowledge of the relevant facts." Additionally, the appraiser must consider the ***most appropriate market*** for an individual item when researching fair market value. This is understood to be the market in which the greatest number of sales (for items like that being researched) between willing buyers and willing sellers is found.	**Estate** is the umbrella term used to describe numerous markets in the jewelry industry. Since an estate appraisal is utilized under different circumstances, it is well to identify the action (Divorce, etc.) if the label *estate* is used. Prices noted in such appraisals are estimates of current Fair Market Values of the items in their present condition since used-jewelry prices do not represent replacement value of new merchandise; IRS regulations apply. *Divorce* or dissolution of property appraisals are often requested either by the parties involved or their attorneys to determine the fair market value of jewelry. Valuation is based on fair market value based on research of the most common and appropriate market of the individual items. This may include items bought for investment during the marriage and held as common property. (Not all states use FMV.) *Probate* or inheritance tax appraisals are made in the course of administering an estate for inheritance tax purposes or for equitable division of property among the heirs (see IRS publication 448). *Donation* appraisal is undertaken when jewelry or gemstones are donated to a charity, institution, museum, or other public institution; IRS regulations apply. These IRS publications should be consulted: Forms 561, 526, 8283. **Auction** appraisals are used by both buyers and sellers of goods at auction for determination of quality and current fair market value of items. **Antique** appraisals take into consideration that antique jewelry (items 100 or more years old) cannot be exactly replaced in today's market; therefore, value based on estimated costs to replace the item in newly manufactured condition would not be appropriate. Valuation estimates for antique jewelry, therefore, must be a reflection of the sum to replace with an item of similar condition, motif, and degree of workmanship, subject to the availability of a like article in the current marketplace. Occasionally, an appraisal of antique jewelry may be undertaken to authenticate the item. Research for authentication includes provenance, attribution, and identification.

Table 1-1. Purpose and Function of Appraisals (*continued*)

Purpose	Function
Fair Market Value (*continued*)	**Collateral** appraisals use fair market value to determine the cash value of jewelry that may be used for a loan request or tendered in lieu of cash for repayment of a loan. **Casualty Loss** The IRS will allow an income tax deduction for lost items not recoverable by insurance, based upon documentation that establishes values for the lost articles. The IRS requirements state that the value be based on the original cost or the current market price, whichever is lower.
Liquidation appraisal is requested by an owner who wishes to convert assets to cash. This valuation is normally lower than replacement value and generally below wholesale value. The function of the liquidation could be for distress sale (now), forced sale (soon), quick liquidation (2 months), orderly liquidation (6 months), or regulated liquidation by IRS (30 days) or FDIC (90 days).	**Bankruptcy** proceedings are governed by the federal Bankruptcy Act (11 U.S.C.A.) and Official Rules and Forms. Straight bankruptcy is in the nature of a liquidation proceeding. Therefore, the appraiser who accepts an assignment of this nature looks for conversion of jewelry to an immediate cash value. In some circumstances, the court may request an ordered liquidation with specific time constraints for liquidating, usually 3 to 6 months.
Other values may be found on various market levels that make up the spectrum of markets. For instance, the appraiser may be asked to research prices at mine sites, cutting centers, dealer-to-dealer, dealer-to-public, or dealers who routinely sell to other dealers and the public at traveling gem and mineral shows.	**Wholesale** value is usually thought of as that price retail jewelers expect to pay for articles they will resell. Such an appraisal might be commissioned by banks and/or attorneys. **Scrap** (melt) is the value of the materials and components of an item, for example, the gold in a 14K ring.

CHAPTER 2

APPRAISAL CONCEPTS AND PRINCIPLES

Valuation Approaches

The Cost Approach and the Market Data Comparison Approach are the two most viable means of estimating value in gems and jewelry appraisals. One of these approaches should be used when the appraiser is researching and estimating value in any category of jewelry. The principles are not only generic to the appraisal profession, in general, but to the personal property branch of appraising in particular. The Income Approach is used in appraising income-producing property (such as leased jewelry) and seldom applies to personal property, but it is regularly used in real property and business valuation. It is not necessary for the jewelry appraiser to learn this method, but be aware of its existence and the appraisal discipline where its use is most appropriate.

Cost Approach

The cost approach is the method most often used to determine retail replacement values, new, for insurance. This method is especially effective with machine-made, mass-produced, simple style jewelry, as well as the basic commercial styles such as diamond cluster rings and mother's rings with multicolored gemstones. The cost approach is *never* used to determine fair market value, nor is it used to estimate the replacement value in antique or designer jewelry. The cost approach requires the valuer to figure the cost of individual components in a piece of jewelry (metal, gems, setting, labor, and so on), and then factor in an appropriate retail markup to obtain the *current retail replacement value* prevalent in the appraiser's locale. Cost approach is used to find the cost necessary to replace—or replicate—a jewelry item, *new*, with new materials of like kind and quality and at prices current in the market at the time of the appraisal.

How to Use the Cost Approach Method

The cost approach analysis is based upon the objective theory of valuation science that says no one will pay more for a property than it would cost to produce a reasonable substitute (the Principle of Substitution). A cost analysis estimates the amount of money necessary to reproduce the subject property as of the report date of the appraisal. The cost new *must* be the cost of replacing or reproducing a duplicate item of jewelry in the current market.

Using the cost approach requires researching and tabulating the price of individual components in an item of jewelry, adding in charges for labor and setting of gemstones, and factoring in a markup to arrive at a final sum as the estimate of value. There are ten basic procedures to follow in using the cost approach, and they are used in exactly the same series of steps on every item of jewelry appraised by this method:

1. Clean, assay, weigh, and measure the jewelry.
2. Analyze the construction.
3. Analyze quality of design and craftsmanship.
4. Figure metal price, design time, labor finishing time.
5. Quality-grade any gemstones using internationally known nomenclature such as Gemological Institute of America terms.
6. Determine approximate weight of gemstones.
7. Price the gemstones.
8. Determine cost of heads, prongs, and so on, used for settings; determine the cost of labor to set.
9. Add up the individual component prices of the metal, labor, and gemstones.
10. Factor in the average markup (used in the appraiser's region) to arrive at a final estimate of retail replacement value.

Occasionally you will examine a mounting with a well-defined maker's mark and karat gold stamp. Sometimes the mounting may be marked with a manufacturer's model number; in this case, use the information to get the price of a duplicate from the manufacturer instead of pursuing the cost approach method to determine the price of the mounting. This is most effective if you are looking for retail replacement value for an insurance appraisal.

Step 1: Clean, Assay, Weigh, and Measure

Cleaning. Most jewelry will have to be cleaned to be appraised. Care must be taken that loose or broken stones are not dislodged or lost from their settings during cleaning. Steam cleaning and ultrasonic cleaning is discouraged if the item of jewelry contains any gemstones under the hardness of 9 on the Mohs' scale. Also, jade, lapis lazuli, turquoise, opal, coral, or pearl should *never* be put into a commercial cleaner of any kind (the Appendix has a chart of complete care and safe handling of gemstones). Diamonds are the exception to the above statement and can *generally* be put into a steam cleaner or ultrasonic cleaner without incident. Of course, a soft brush, a mild soapy water solution, and plain water rinse treatment is safe in nearly all instances. If a diamond ring is dipped into a small cup or bowl of rubbing alcohol, it will remove any remaining soap scum from the diamond and produce a nice sparkle.

Assaying. Metal type and fineness must be determined before any estimate of value can be obtained. It should be standard practice for the appraiser to test all mountings with test acids even if the item is stamped for metal fineness. Results should be recorded on a worksheet. Test kits, test needles, and rubstones that provisionally test metals are available from all jewelers' supply houses. Some are prepackaged, as shown in figure 2-1, in wooden boxes. The acid testing procedure is easy, but practice is necessary until the appraiser is confident with the results achieved. Certain nondestructive ways of estimating gold content, sight, and heft, cannot be trusted as accurate. Some appraisers occasionally use specific gravity testing, but this can be misleading if there is a large volume of stones in the piece or a combination of metals.

Specific Gravity:		
	Platinum	21.45
	Gold	19.32
	Silver	10.50
	Nickel	9.81
	Copper	8.94

It is a United States Marking and Stamping law that if a fineness mark is present, the piece must bear a manufacturer's trademark—an identifiable symbol or initials.

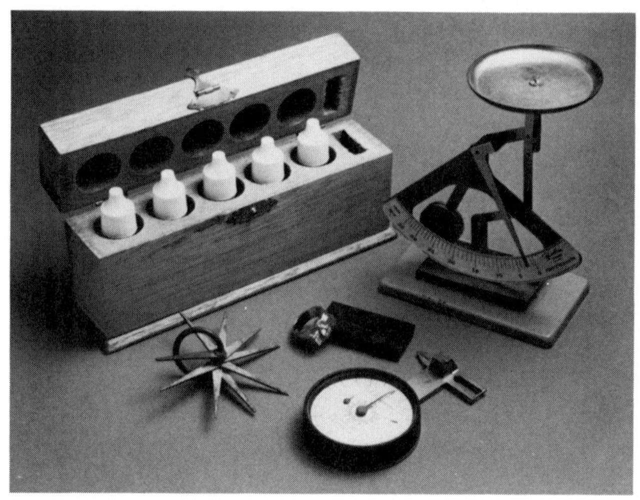

2–1. Metal testing kits, scales, and gauges are used for accurate evaluation.

Steffan Aletti, president of the Jewelry Industry Council, has warned, "It is illegal to sell jewelry that has a fineness stamp but no trademark. The only exception is when the piece is too small or delicate to be stamped—a light, filigree earring, for example." In those cases when an appraiser finds no trademark or maker's mark on the item of jewelry, it should be noted on the appraisal report that no manufacturer's name is present. There are no laws that mandate a fineness mark.

Gold jewelry sold in the United States must be a minimum of 10K to be designated by the word *karat.* Platinum is marked by its full name or by *plat.* Sterling silver is marked by one of three designations: *STERL., Sterling,* or *.925.* The silver alloy designated *.900* has less silver than sterling and is known as coin silver. Gold filled must be a minimum of 1/20th (5%) gold; rolled gold plate and gold overlay are 1/40th gold.

In its instruction on testing metals, the Gemological Institute of America warns that information gained by testing metals with nitric and hydrochloric acids is only an *indication* that the quality of the metal is as stamped. Fire assaying, according to GIA instructors, is the *only positive method* of determining the identification and quality of an unknown metal. They warn, however, that this is a destructive test since, to get positive results, a small section of the unknown metal must be removed from the jewelry item.

Besides using a basalt stone or rubstone for streak testing, many experienced appraisers use a test that can—and often does—mar a piece of jewelry. This test must be used with great care and is not recommended as a general practice; a file, emery board, or sharp metal point is used on an inconspicuous spot to secure a small mark. Then, a reduced nitric acid solution (5 parts nitric acid to 7 parts distilled water) is dropped on the mark and the reaction of the mark is observed. Low karat gold will have a slow, green, effervescing reaction; high karat gold will not react as quickly. A violent green reaction identifies the underlying base metal and indicates a gold-plated or gold-filled item.

Weighing. The weight of the jewelry can be described either in grams or pennyweights (dwts.), and the weight of the metal should be reported on the worksheet and later on the finished report. If the jewelry contains gemstones, the appraiser can use the following formula to estimate weight of the *metal only:*

- *Grams:* Obtain the total weight of the item in grams. Convert carat weight of the gemstones into grams using this formula:

$$\text{carats} \times 0.2 = \text{grams}$$

- Deduct the gemstone gram weight from total gram weight of the item for a final estimated gross weight of the metal.
- *Pennyweights (dwts.):* Figure the net weight in pennyweights of the mounting, including any set gemstones, and then use the following formula:

$$\text{net weight} = \text{gross weight minus weight of the stone in carats} \times .13$$

- If the identity of a stone set into a mounting is unknown, use an estimated specific gravity (S.G.) of *3.00* to figure approximate carat weight.

Measuring. For more thorough measurements, note widths of jewelry items (especially ring shanks and graduated width chains). Report the taper or graduated range of width from smallest to largest point in the appraisal description. Also, note the finger size of rings to insure the most complete description and measurement.

Measuring jewelry articles cannot be overemphasized. This is an important appraisal step for identification and calculating value. A Leveridge gauge, millimeter gauge, standard measuring tape, and any other measuring equipment, such as gemstone table gauges, should be conscientiously used for all measurements; and the results should be carefully recorded. There is a standard way to measure and record, with the general rule being vertical before horizontal. Usually, the dimensions of height, length, width, depth, or diameter (if all are applicable) are taken at the greatest point. If this is not possible, a notation should be made on the appraisal document explaining why the article was measured in an atypical way. Height always precedes any other measurement record. Depending upon the article, the appraiser may wish to record separate measurements for the component parts, if such exist.

In general, dimensions are taken with notations that indicate if a clasp, bail, loop, or other attached component parts are included. Both English inches and millimeter gauge measurements should be recorded.

Step 2: Analyze Construction

An appraiser who has knowledge of jewelry construction, or one who has taken jewelry-making classes, will be ahead of his or her peers in the ability to discern and interpret manufacturing methods, techniques, and degrees of craftsmanship. This knowledge is important because some types of jewelry manufacture are more labor-intensive; thus the final product is more expensive. Mountings that are mass-produced by casting and tumble finishing can be sold for far less than those finished by hand or with a combination of cast and hand assembly.

Types of Manufacturing. Cast manufacturing is the most popular method used in the United States today by commercial jewelry companies. An estimated 95 percent of all karat gold jewelry is of cast manufacture. The reason is that the cast method permits inexpensive replication of design and detail, dispensing with labor-intensive handcrafting. It has soared in popularity since the 1940s, when casting equipment was improved and the price of gold was holding steady at $35 an ounce. Today's state-of-the-art casting equipment has raised this method to one that can produce extraordinarily fine mountings, *if the manufacturer is willing to invest in extra finishing time,* such as the example shown in figure 2-2.

2–2. An example of a well-finished casting.

All cast items of jewelry start with a model made either in wax or metal from which a rubber mold is made. The mold is made in halves that can be taken apart and used again and again. The mold is injected with wax and opened after the wax is set. The carefully removed wax model is fitted with a *sprue* (a channel of wax attached to the model through which the metal will flow), and attached to a *tree,* a wax base that may have as many as two hundred wax models attached at one time. The tree is placed in a flask and filled with an *investment,* a slurry mixture of plaster of Paris, and set aside to harden. After hardening, a *burnout* of the wax takes place in an oven. This step in the procedure can last up to eight hours. After the burnout, a negative cast results with space that replicates the now lost wax model(s). The plaster cast is then placed in a crucible while the metal to be used is brought into the molten state. When the metal is liquid and able to flow, it is flung into the negative space(s) of the wax model with centrifugal force. Once removed, the metal items are cleaned, finished, and polished. One of the by-products of an unsuccessful burnout is residual carbon monoxide that causes surface porosity in the metal article. Particular attention should be paid to porosity, pits, or bubbles seen on the underside of mountings, as in the example in figure 2-3. These are the result of the cast method and are often seen in hastily cast articles. When pitted or weak castings with rough surfaces and flaws in the casting such as folded gold (fig. 2-4) are observed, it is a certain sign the item is a product of the casting process and points to the quality of the manufacture. It also speaks plainly to the appraiser about production cost control.

2-3. Porosity in a cast piece of jewelry.

Die-struck items can sometimes be identified by the bright finish seen in the small areas of the mounting. Once the most profitable way to make jewelry, this method has become, with the advent of better casting equipment and techniques, more expensive than rubber molds for casting. Die striking requires extra space for the machinery and involves a much higher equipment cost. In addition, each design must be produced in quantities of more than 100 to absorb the cost of new hub, die, force, and so on.

In using a die, metal is rolled to a thickness greater than that of the desired item, then sandwiched between a matching punch and die. Industrial dies and punches usually shear the metal as well as shape it. A lot of commercial and costume jewelry is still produced using this process: wedding and engagement ring sets, promise rings, earrings, guard rings, birthstone rings, and so on. Also, many units or separate pieces, such as leaves and flower decorations, are prestamped and assembled on mountings (fig. 2-5)

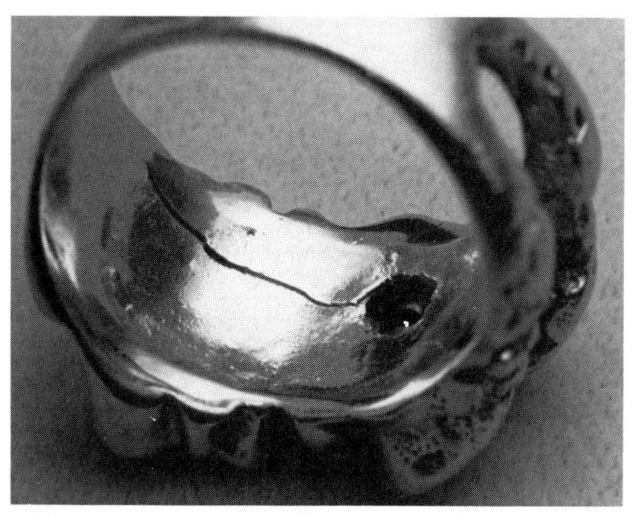

2-4. The folded gold and unfinished underside is a mark of a poor casting job.

2-5. Prestamped flower-and-leaf decoration on a ring.

2–6. Inspect jewelry backs for indications of handcrafting.

2–7. Even handmade jewelry may lack finishing details.

that may themselves be either die-struck or cast and assembled. In figure 2-5, the characteristics of die-stamping can be seen in the similarity of the leaves, indicating they were made in the same die. They were prestamped and then soldered onto the mounting. One sign to the appraiser of the die-struck method is that, while the underside of mountings may look evenly finished, any galleries around the outside edges may look unfinished on the back or underside.

Although it is true that *handmade* jewelry will usually have a higher value than jewelry constructed by other methods, a poorly made piece of jewelry could be less valuable than the same item cast and finished with special attention to detail. Typical indications of handmade jewelry items are: overlapping and sometimes incomplete cuts that do not join other parts of the design (figs. 2-6 and 2-7) and the tool marks. At

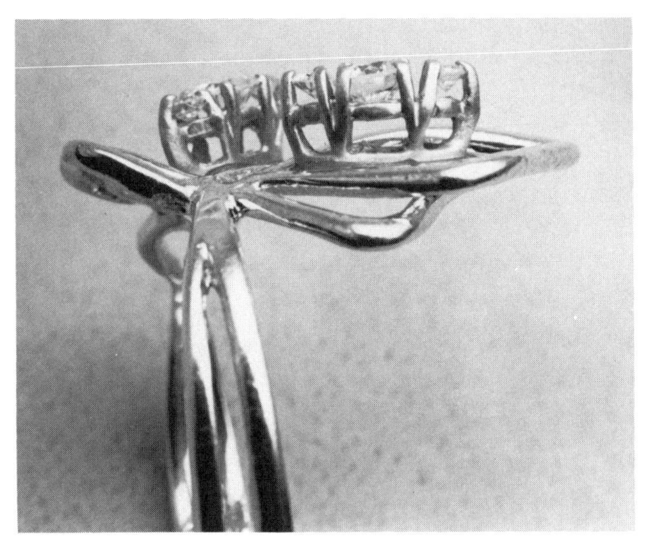

2–8. Assembled jewelry is common practice. The ring shown has a cast shank with settings (heads) soldered onto the top.

times handmade jewelry will show a different color of solder in the joints. Appraisers must be mindful that the Federal Trade Commission (FTC) holds it is unfair trade practice to represent, directly or by implication, that a jewelry item is handmade or handwrought unless the entire shaping and forming of such product from raw materials to finish and decoration has been accomplished by hand labor, without the aid of mechanical devices other than ordinary hand tools. Also, the major components, including joints, settings, and so on, must be handmade from plate, bar, or wire. If *any* part is cast or die-struck, it cannot be called a handmade item! The specific ruling from FTC 16CFR Part 23, February 27, 1979, is:

> 23.4(b) It is an unfair trade practice to represent directly, or by implication, that any industry product is hand forged, hand engraved, hand finished, or hand polished, unless the operation described was accomplished by hand labor and manually controlled methods which permit the maker to control and vary the type, amount, and effect of such operation on each part of each individual product.

A lot of jewelry is a combination of manufacturing methods, especially rings that can have shanks cast or die-struck with tops die-struck or handmade. Also common are cast mountings (all types of jewelry) with heads, or settings (fig. 2-8), die-struck and assembled onto the item.

Mountings and settings are the nomenclature used for fixing stones in pieces of jewelry. A mounting refers strictly to the construction of a metal part of the jewelry including the supports for the stones. It is the *overall* name given to the frame or artistry of the

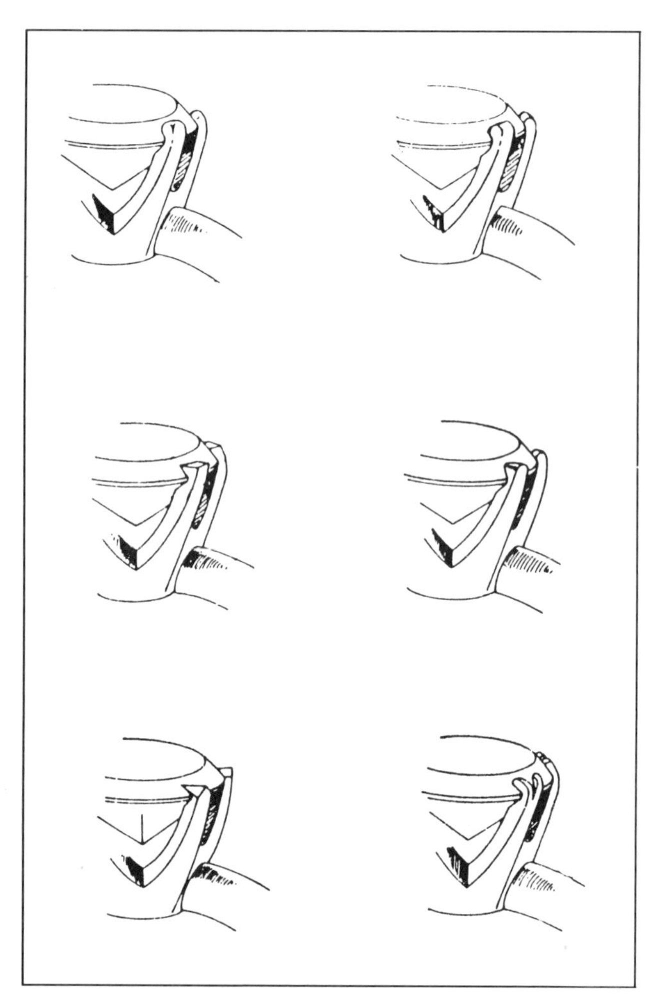

2–9. Prong Styles. *Top left to right:* round or ball, flat-sided round; *Center:* flat top, barrel or quonset; *Bottom:* knife edge, split. (*Reprinted with the permission of the Gemological Institute of America.*)

2–10. Gypsy setting.

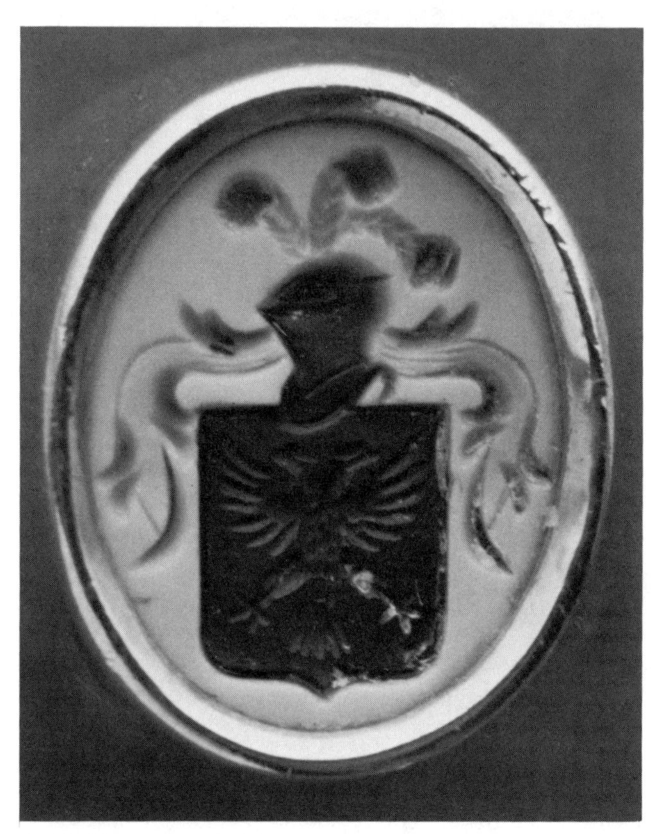

2–11. Roman setting.

metal. Supports for the gemstones are called *settings,* or *heads,* and these too have specific terminology. Mountings and settings are identified as handcrafted, cast, or die-struck. Appraisers must be comfortable with the identification and proper nomenclature of the various types and styles of settings because they have a critical bearing on the accurate description of an item on the appraisal report.

The setting styles most commonly seen are the following:

- *Prong setting* is the most commonly used setting in jewelry. It requires setting a stone in a head or basket setting. Most heads are die-struck and are usually made to fit stones of standard calibrated sizes. Basket settings are usually cast or fabricated from wire and made to fit an individual stone. Prongs are made in a variety of styles as illustrated in figure 2-9: round or ball, flat-sided round, flat top, barrel or quonset, knife edge, and split.
- *Gypsy setting* (fig. 2-10) is used with a *domed* ring mounting and gemstone, usually cabochon cut, surrounded by a tight edge of metal. There is not always a bezel because sometimes the stone is set flush and no prongs are used.
- *Roman setting* (fig.2-11) is often confused with Gypsy setting, but it has a channel around the outside of the stone that makes the stone appear framed. This type of setting was used for Roman seal rings. When the seal was pushed into wax, it looked like it was in a frame and no prongs are used.

2–12. Bezel setting.

2–14. Grain or bead setting (shown in star design).

2–13. Tube setting.

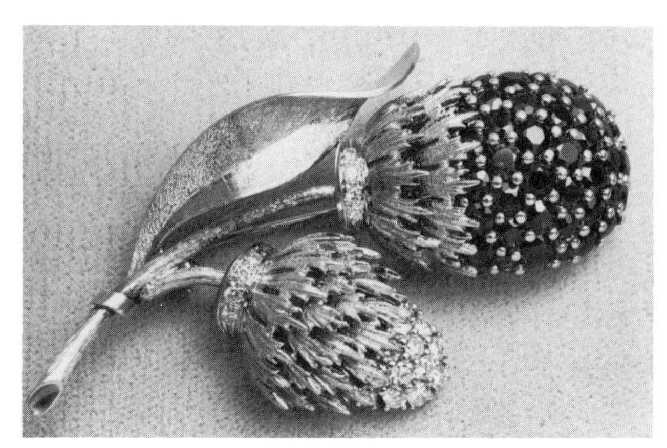

2–15. Pavé setting.

- *Bezel setting* (fig. 2-12) is use of a rim of metal shaped to fit around the girdle of a stone and then soldered onto a mounting. Cast bezels are also available with open and closed backs. This is also known as *Collet setting* and sometimes *Pronged Collet setting.* A contemporary outgrowth of the bezel or collet is the
- *Tube setting* (fig. 2-13); a straight or tapered tube instead of a rim is used to surround the gemstone. Both Tube and Collet settings have also been known as *Chenier settings.*
- *Grain or Bead setting* (fig. 2-14) involves using tiny grains or beads from the metal surface to secure the stone. This is often used with other styles of settings such as prongs or bezels. In this method, stones are arranged on the mounting, seats are burred and beads of metal are raised over the edges of the stones' girdles onto the edge of the stone.
- *Pavé setting* (fig. 2-15) is a classic example of using the bead technique and may be employed in two ways. Sometimes the gemstones are all the same size and are lined up girdle-to-girdle in rows. When they are of different sizes, they are fitted into a given area of metal in random fashion.
- *Channel setting* (fig. 2-16, see page 14) is a contemporary style. Stones are set deep in the mountings, which protects the girdle. When this technique is employed, the stones are seated between two walls of metal in a channel. Stones can be pressed or slid into place.
- *Tension setting* (fig. 2-17, see page 14) is a technique whereby stones are held in place by the tension of the metal mounting.

2–16. Channel setting.

2–18. Fishtail setting.

2–17. Tension setting. Ring by José Hess (*Photograph courtesy of Diamond Information Center*).

2–19. Cluster illusion setting.

- *Fishtail setting* (fig. 2-18) has a fishtail formed by curving the holding prongs toward each other. The curving of the metal to the side leaves a notch in the center and gives the appearance of a fish tail.
- *Cluster Illusion setting* (fig. 2-19) uses a group of stones held by prongs and arranged to give the illusion of a single large stone.
- *Illusion setting* (fig. 2-20) is the term used to refer to heads that enhance the appearance of small stones, making them seem bigger. Metal reflections around the stone add an illusion of size.
- *Basket setting* (fig. 2-21) is cast or fabricated from wire to fit a particular stone. It can also be assembled onto a mounting and used with prongs in any number of styles.

Step 3: Analyze Quality of Workmanship and Design

There are several points to consider when examining an item for quality of craftsmanship. One of the most obvious but overlooked clues to well-finished jewelry (thus more expensive) is finish—or lack of—in the settings. Observe the surface inside and under-

2–20. Illusion setting.

2–21. Basket setting.

neath the prongs and inside channels. Handmade pieces should be polished and glowing like a jewel. Cast pieces showing care of finishing will reveal polish underneath the settings and in between metal levels of a mounting. This tells you that the finisher took trouble with a needle file, smooth file, brush, and possibly even used the thrumming technique to get at hard-to-polish areas. The thrumming method involves the use of a length of soft cotton string or cord saturated in a slurry of polishing compound and water. The string is then drawn back and forth over the area to be polished. Because thrumming is labor-intensive, it is seldom employed on less expensive goods. However, if you are using a qualitative ranking system, that is, good, better, best, to help in the determination of value, workmanship will signal a ranking position and offer the most information.

If an item is well-made and finished, or poorly made and finished, it impacts on the final estimate of value. Careful scrutiny by the examiner is necessary so that he or she can explain on the appraisal report or at some time in the future, if necessary, the criteria used in estimating value. A chief observation should be the care afforded to the gemstones. How securely are they set? Is the mounting and setting proportioned to the stone? How much wear is on the stones and settings, and what is their condition? Alison Birch LeBaron, Master Gemologist Appraiser of GME Jewelers in California, makes these observations when doing a jewelry appraisal: "I check to see what repairs may have been done, or needed. The stones are the focal point of most jewelry items; it is vital that the stones are displayed well and safely." She also adds these points about checking mountings, "Is the mounting even (as in shanks of rings); do the curves flow smoothly? Are the lines straight and angles equal? Is the piece balanced?"

Alan Revere, owner of the Revere Academy of Jewelry Arts in San Francisco and an award-winning jewelry designer/craftsman, always looks at the backs of jewelry items first. "The back is much more revealing than the front," he insists. "Look to see if the mounting has pits and if the lines are clean and straight as these are the most revealing signs of the level of craftsmanship." He adds that a ring with a shank or gallery design important to the integrity of the item should have those designs well-executed, with all file marks polished out and any pierced or filigree gallery work smooth. Sloppy work is often revealed by an unfinished or semifinished underside (fig. 2-22). "To find a burr of metal inside a shank certainly speaks negatively to the workmanship—and thus to the value of the piece."

Handmade mountings are given the greatest scrutiny and deservedly so, for fine handmade mountings will be awarded a premium value over any other type of manufacture. Revere insists that in handmade mountings the backs should be as beautiful as the fronts, with the areas *behind* the gemstones completely finished and polished. "This is where the craftsman's care and competence shows up," he says. Another important consideration to the quality of the mounting is a trademark.

2–22. Unfinished gallery underside marks hasty work.

2–23. Both karat mark and maker's mark should be clearly stamped inside the ring shank. In this ring the maker's mark is missing. The letters *OB* stand for outer band; and this ring is marked 10K gold.

2–24. Stampings make jewelry assessment easier and give a starting point for confirming metal fineness.

The trademark, or maker's mark, and karat mark should be clearly and evenly stamped (figs. 2-23 and 2-24). Occasionally, an article will be found with a maker's mark stamped on a small plate and soldered to the mounting. Many Tiffany articles are stamped in this manner. "This," Revere says, "is a plus for the quality of the mounting."

Other indications of well-made mountings are the azured backs, either by machine as in figure 2-25 or by hand as in figure 2-26. This technique is always found on finer goods. When hand-azuring is found, the appraiser can be confident in adding at least 30 percent to the cost of the labor, according to Robert Sandler, owner of Designer Jewels in Houston. "The appraiser should recognize that there *is* a difference in values because of the application of this technique."

Replicas. A more difficult identification to master is the separation of genuine and knock-off antique and period jewelry articles. Many highly skilled craftsmen are able to replicate known designers' work as well as period and modern jewelry. Much of this jewelry comes into the United States from Asia and the

2–25. Example of a machine-azured setting.

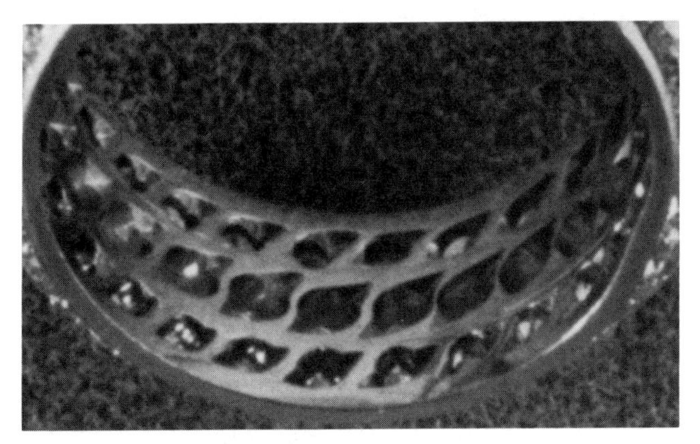

2–26. Hand azuring indicates care and quality of workmanship. (*Photograph courtesy of Designer Jewels*).

Pacific-rim countries. Only vigilance, examination, research, and investigation can protect the appraiser's integrity, analysis, and value estimates.

Occasionally there will be a piece of jewelry totally lacking in balance and symmetry as well as finishing, which are essential to a well-priced work, but it may have a well-known artist or designer's signature. Investigate carefully the possibility of a replicated piece before you attribute it to somebody famous. The craftsman duplicating it may have been working from the memory of another designer's work, or amended some details to achieve a price. Each and every piece must be appraised on its own merits, even if the famous designer's signature turns out to be authentic. Remember that even some well-known designers and artists have had bad days and bad designs. The bottom line is that while a piece may be attributable to a famous designer, and his or her name will raise the value somewhat, the item cannot be judged simply upon the past successes or reputation of the designer. The item is subject to the same quality grading as any other jewelry article. Placing value on one-of-a-kind items requires logic and use of comparison item value.

Step 4: Price the Metal, Design, Time and Labor for Finishing

A basic formula for computing the cost of gold, platinum, silver, or other metals is used by determining the daily London spot price of these metals. The spot price may be obtained by telephone from a precious metals dealer or from the daily newspaper under the metals spot prices.

The basic formula is as follows:

1. Obtain the current gold spot price (for example, $450 per troy ounce).
2. Determine the fineness wanted (for example, 14K).
3. There are twenty (20) pennyweights in an ounce of fine gold; therefore, to find the cost of 14K gold per dwt. at the current gold price of $450 per oz., apply the following:

$$\frac{\$450}{480} \times 14 = \$13.13 \text{ (per dwt.)}$$

The constant 480 figure is obtained by multiplying 20 dwts. (to the ounce of gold) times 24, the karatage of pure gold. To find the price per dwt. of 18K, 22K, or any other gold fineness use the same formula, but multiply by the alloy wanted. To obtain the intrinsic metal value of a 5 dwt. 14K gold wedding band, for example, use the formula and multiply the per-dwt. price of gold and the dwt. of the jewelry item.

Example:

$$\frac{\$450}{480} \times 14 = \$13.13 \text{ (per dwt.)} \times 5 \text{ (dwts.)} = \$65.65$$

To estimate the manufacturer's selling price of an item (for example, a 14K gold 5 dwt. ring), use the average ratio markup of 1.5 to 3.0 times the gold content. This figure will vary with the complexity or simplicity of the ring, the individual manufacturer, and the geographic region. All of this is more reason the appraiser must be prepared to do research and legwork in the local marketplace to build a personal and comprehensive pricing table. The average cast or diestruck ring will be marked up about 1.5 x by the manufacturer. For a one-of-a-kind ring, allow 2.5 x to 3.5 x to cover design. Intricate rings and tricolor gold rings will be marked up at least 2.5 x. Platinum rings will be marked up about 3 x because the work in manufacturing and finishing platinum is more labor-intensive than in gold rings. After determining the manufacturer's markup, the appraiser can deduce the estimated wholesale figure of the item (unadorned and without gemstones) as it is probably sold to the retail jeweler.

Steps 5, 6, and 7: Analyze and Price Diamonds and Colored Gemstones

It is beyond the scope of this book to give a course of instruction in diamond and colored gemstone identification and quality grading. Individuals who aspire to appraise gem-set jewelry must have gemological training. It is impossible to value an object without rudimentary knowledge of the item. Indeed, it could be likened to a practicing surgeon skipping the lessons on human anatomy. *Basic gemological schooling is fundamental to the practice of jewelry appraising.* It should be noted, however, that when diamonds and colored gemstones are part of the jewelry to be appraised, a complete examination and appraisal report should provide the following data.

2-27. The correlation and analysis of prices for gemstones and jewelry is a major step in using the Cost Approach method to estimate value.

Diamonds:

- The carat weight with millimeter dimensions in length, width, and depth should be shown. If the depth cannot be taken accurately because of the design of the mounting or setting, it should be explained on the report along with the estimated depth. Record the shape of the diamond.
- Color grade of the diamond should be noted along with a notation of the grading system used, such as GIA, American Gem Society, or any other. There should also be a notation if master diamonds were used to grade for color comparison.
- Note the clarity grade with mention of the grading system used, and the type and power of magnification used to judge clarity.
- Comments should be noted on the quality of the proportioning and finish of the stone(s).
- The fluorescence of the diamond should be discussed along with comments on the reaction of the diamond under both ultraviolet longwave and shortwave examination.
- A photograph of the diamond jewelry and a plot of the diamond (for one-carat-and-over sizes) should be included in the appraisal document.

Colored Gemstones:

- On the appraisal document, list both the species and variety of the gemstone, the shape of the gemstone(s), and comments as to whether the gem is natural, synthetic, simulated, or a composite, plus the quality grade of the material.
- Size (whether loose or mounted): The carat weight should be estimated using Leveridge gauge millimeter measurements (length, width, depth) applied to a standard weight estimation formula. Note the measurements and use of weight estimation formula in the report.
- Color of the gemstone should be noted along with remarks about the hue, tone, and intensity of the color. The color grading system used should be indicated. If the color is natural, it should be stated. If the color is due to heat, irradiation, or other treatment, this should be made clear along with a notation of the instrument(s) used to confirm the observation.
- Record fluorescence under ultraviolet.
- Include a photograph of the item in the report.

Fundamental to appraising is the gathering and maintenance of diamond and colored gemstone prices (fig. 2-27). Diamond prices change rapidly, sometimes weekly or monthly, depending upon the strength of supply and demand. Colored gemstone prices fluctuate in a slower mode. Along with self-maintained pricing, monitoring dealers and wholesale suppliers and attendance at gem and jewelry shows, as well as subscribing to price guides, will help bring some consistency and the means of document substantiation to market research. The knowledge of the lowest price a stone dealer will accept, and not merely the dealer's asking price, is essential information.

Steps 8, 9, 10: Determine Costs of Settings and Labor, and Add Retail Markups

The prices listed here are to be used for comparison purposes only. They will change from region to region according to local economics. Some *average* trade shop jewelry setting prices in Houston, Texas, 1988–1989 (Keystone) are:

Diamond setting:

Melee-prong-needlepoint	
Up to 4 points	$ 7
5 to 12 points	9

Flat plate, bead, bright cut	
Up to 4 points	$ 9
4 to 12 points	11

Ladies' center prong or needlepoint

Up to 12 points	$ 9
12 to 24 points	11
25 to 49 points	13
50 to 74 points	18
75 points to 1 carat	22

Man's center flat plate, bead, and bright cut

Up to 10 points	$14
10 to 24 points	17
25 to 49 points	23
50 to 74 points	28
75 points to 1 carat	33

Center heads, illusion or prong

0 to 10 points	$40
1 to 25 points	45
6 to 50 points	50
1 to 75 points	60
75 to 1 carat	70

Ladies' 4 or 6 prong setting, includes setting

Up to 5 points	$25
6 to 10 points	30
11 to 21 points	35
22 to 50 points	40
51 to 75 points	50
76 to 1 carat	60

Man's flat plate, bead set including setting

Up to 12 points	$35
13 to 25 points	40
26 to 50 points	45
51 to 75 points	55
75 points to 1 carat	60

Claw and Bezel setting

5x7mm, 6x8mm, 8x10mm	$12
10x12mm, 12x14mm	14
12x16mm, 14x16mm	16
13x18mm	18
15x20mm	20

Bezels for gemstone rings including setting

8x10mm	$45
10x12mm	50
12x14mm	55
12x16mm	60

Average Retail Markups

The following average markups are compiled from surveys of jewelers nationwide. Caution is advised, however, in zealous use of this table because markups vary from one geographical locale to another. The individual jeweler's markup is dictated by one's buying power, terms of sales, and overhead. Nevertheless, a consistent markup can be found in most communities, and the following schedule can be viewed as a reasonable guide.

Cost	Markup Times Cost
Up to $300	3 x
$301 to $500	2.50 x
$501 to $2,500	2.00 x
$2,501 to $5,000	1.90 x
$5,001 to $10,000	1.85 x
$10,001 to $20,000	1.75 x
$20,001 to $50,000	1.60 x
$50,001 and over	1.50 x

Using All the Steps in the Cost Approach Method

The ring (fig. 2-28) is the subject used in a recapitulation of the Cost Approach Method.

- *Steps 1 to 4:* 14K yellow gold die-struck mounting set with a three-carat marquise-cut diamond, F color, VVS-1 clarity, in a 6-prong 14K white gold head. The mounting weighs 6 dwts.
- Current spot gold price: $401.
- Gold per pennyweight: $11.70.
- Intrinsic gold value of the ring: $70.20.
- Average manufacturer's markup results in $140.40 wholesaler to retailer price for mounting.
- Price to consumer approximately $351.00.
- *Steps 5 to 9:* Current wholesale price of three-carat diamond, F color, VVS-1, $11,500/ct: wholesale to retailer $34,500. Price to consumer: $55,200. White gold 6-prong head, labor and setting: $180.00.
- Add up consumer cost of mounting, diamond, and setting to get a final figure of $55,731.00, the appraised value of the diamond ring, without considering any added taxes.

The $55,731 figure represents the average current 1989 retail replacement value (new) of a ring similar to that illustrated, using typical markets and average markups; however, appraisers should rely on markups they know are used in their own locales.

2–28. Three-carat marquise diamond solitaire ring. (*Photograph courtesy of Diamond Information Center*).

Market Data Comparison Approach

Using this value technique requires the appraiser to determine the highest-and-best use of an item in the most common market for the jewelry being appraised, and to research and be able to substantiate the valuation figure with trackable sales of similar or identical properties. This approach is used when the appraiser seeks to establish fair market value for antique or heirloom jewelry, for divorce, donation, collateral loan, casualty loss, and other estate appraisal matters. It can be, and often is, used for insurance retail replacement value when a piece is of fine designer quality, or stamped by Cartier, Tiffany, or another illustrious firm such as Bulgari or Van Cleef & Arpels. When used for insurance, the item being appraised is compared to either a new duplicate from the same manufacturer and designer or with a comparable item found in the secondary jewelry market. The item must be clearly labeled as being *new* if the appraisal is being sought for insurance, or, if not a new item, a notation must be made that the value is for a comparable piece. Often a comparable is found that is not attributable to the manufacturer of the jewelry item being appraised but is identical and has the same characteristics as the subject item. This is acceptable, but all this information should be fully stated by the appraiser in the report. In some instances, a client will request an *exact* duplicate by the *same designer* for replacement in case of loss. That will necessitate the appraiser's call or letter to the company or designer to obtain the price of an identical item. In the case of well-known designers or European-manufactured jewelry, this research can involve considerable time spent on the client's behalf. The appraiser deserves compensation according to the complexity and time involved in the research.

The Market Data Comparison Approach is straightforward but often misunderstood. It involves the actual gathering of information about historical sales of similar properties. The report should describe the salient characteristics of the jewelry, and identify any specific and unusual conditions that must be measured in this approach, for example, if the current precious metals price is at an all-time high or low. It is also obligatory to obtain *recent* sales reports of similar properties that will validate the final estimate of value.

How to Use the Market Data Comparison Approach

One sale does not constitute a market price or value. The appraiser who uses comparables is looking at the quantity, quality, and substantiating sales data for several comparables. In this approach, the appraiser must research the most common and appropriate market in the region for similar objects to find sales data that confirms and qualifies the final value estimate. This approach involves four steps:

1. Researching the market
2. Analyzing each of the comparable properties to determine the attributes of each that correspond to ones of the subject property
3. Organizing the information from step 2 and arriving at a plus or minus adjustment for selected comparables
4. Reviewing available data, correlating the data, and developing a value claim.

It is important to confirm and qualify prices for comparables, and review the sales terms involved in the transactions. Although difficult, the appraiser should try and determine events surrounding the sales. Was it for a lump sum? If at auction, what kind of buyer fees or premiums were involved? Was the piece on consignment? Various kinds of sales terms impact the final estimates of value, and the appraiser should at least be aware of this fact.

While world markets are important in finding the value of an object, the appraiser will establish the mode of prices in the locale that most reliably represents the *justified* price, which a typical, informed, rational purchaser would pay for such an item of jewelry.

Comparable sales data should, ideally, not be older than 12 to 18 months, but depending upon the object being researched, the valuer may be compelled to use older sales data.

Valuation sciences teaches a mode, a median, and a mean. If a mode cannot be established in one's marketplace, then the median should be used.

- *Mode*—The mode is the most frequently occurring number in a series of numbers, such as sales transactions.
- *Median*—The median divides a list of numbers into equal halves arranged in numerical order. If the list of numbers is odd, the median is the middle number. If the list of numbers is even, add the two central numbers together and divide by two to calculate the median.
- *Mean*—The mean is found by adding together the sales figures one may have found for an item and dividing by the number of transactions. For example, for ten sales, total all the figures and then divide by ten to get the mean, or average.

These are ways to measure the *central tendency* and the type of information you may need to substantiate your value claims in a court of law.

2-29. Diamond ring comparables. Adjustments must be made for differences of value estimation.

What Is Comparable?

Confusion often surrounds the question, "What is comparable?" If the appraiser remembers to compare *comparable* items and to adjust for those items that are comparable, but not identical, the process can be mastered without undue stress. There are at least eight categories of comparison:

1. Uniqueness
2. Rarity (How many exist?)
3. Quality
4. Condition
5. Origin and Manufacture
6. Quality of design execution or craftsmanship (Is the overall size comparable?)
7. Appropriate market (based on clients purpose)
8. Appraisal purpose (Retail replacement? Liquidation? Fair Market Value?)

In judging comparables, consider the number of sales, the period of time those sales cover, the state of the market during that time (was it unduly influenced by world economic events?), and the degree of comparability. The comparables can be equivalent or identical, but do not have to be either if the elements or attributes are at least comparable. There is no set number of comparables to utilize, but the appraiser will need to have at least three to discern some type of pattern and analyze and arrive at a value claim. The comparables used for value determination should be fully described on the appraisal report if the item is of significance. If not appended to the report, the data should be filed in the client's folder for future reference. Bear in mind that sufficient and appropriate data is one of the biggest supports to competency. A poor appraisal is one that relies on few statistics to reach a conclusion.

The three photographs, figures 2-29, 2-30, and 2-31, show comparable items utilizing measurements of gold fineness, weight of gold, type and carat weight of precious gemstones, condition, quality, appraisal purpose, and supply and demand. A reminder should be added that appearances can be deceiving, especially if selecting comparables from catalogs or sales brochures. Be certain you are comparing like kind!

In analyzing figure 2-29, observe the three diamond rings first to determine if their attributes are comparable. Let us assume that the piece to be valued is the ring in the middle. We must consider the differences and similarities between the other rings sold, along with the date of sale. Unless the sale took place recently, say within the last six months, the appraiser should consider the changes in the market that may have occurred since the sale date.

Analysis of these rings tells the appraiser they were probably produced in three separate decades. The ring on the left is the oldest and can be circa dated 1910, as is evident by the old European-cut diamonds and style of the mounting and settings. Scrutiny of the ring's workmanship shows that care was taken on the shank and shoulders, and they are embellished and well-finished. The subject ring in the center is circa 1945, in excellent condition, with very fine quality diamonds. The ring on the right, circa 1935, also contains fine quality full-cut round brilliant diamonds. All rings have similarly sized center stones. The total diamond weights, gold weights, condition, and workmanship are comparable. Slight price adjustments would be necessary for cutting style of the major diamonds.

The three Egyptian cartouche pendants (fig. 2-30, see page 22) may look the same, but some differences are apparent upon inspection. The three are all handmade from Egypt. The center pendant is 22K, the others, 18K. The center cartouche is heavier (6 dwts., 1½ inches long) and longer than its counterparts. The cartouche at left (2 dwts., $1^{7}/_{16}$ inches long) is light weight. On the right (3.95 dwts., $1^{5}/_{16}$ inches), the

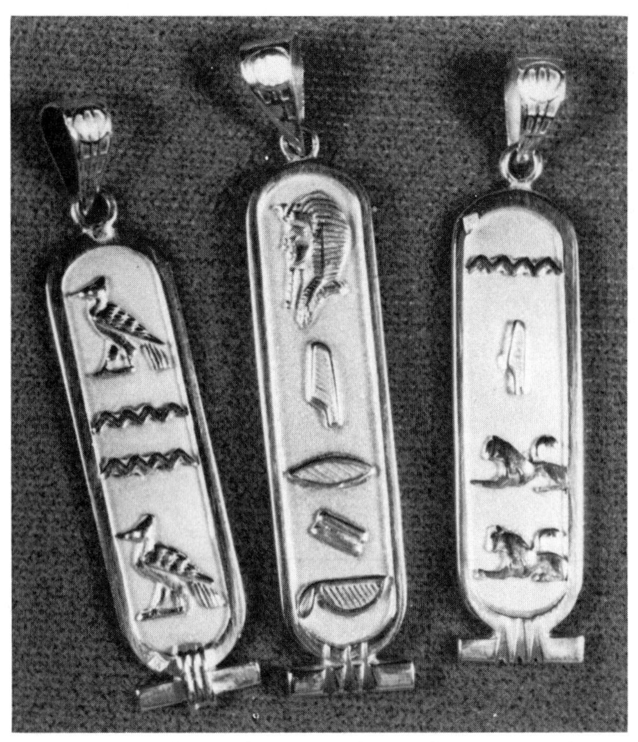

2–30. Identical Egyptian cartouche pendants, or are they?

cartouche is almost twice as heavy and has hieroglyphics on both sides. Approximate retail value of the three in the United States would be: *left,* $200 to $250; *center,* $500 to $600; *right,* $375 to $400.

The three 14K yellow gold bracelets (fig. 2-31) are all from the same company, but they are not good comparables because of the disparity in weight, length, and design.

Finding Information

Where can you gather data about the items you are seeking to value? This is the lament of the appraiser who needs help with identification, especially for antique and period jewelry. Some price information comes from guides available to subscribing appraisers and sales reports of similar items from retail jewelry stores. Gathering additional information requires creative investigation that can begin by referring to this list:

- Wholesale jewelry firms
- Traveling gemstone and jewelry sales representatives
- Manufacturers
- Auction houses (major and local)
- Dealers with specialties in various markets
- Antique jewelry magazines
- Jewelry books and jewelry trade shows
- Gems and mineral shows (both commercial and hobby)
- Jewelry industry trade magazines and catalogs
- Telecommunication networks
- FAX communication inquiries
- Computer information services
- Personal records of past appraisals as well as sales, advertisements, and classifieds in local newspapers
- Peers and colleagues
- Provenance from the owner of the item
- Exhibition catalogs from museums
- Retail jewelers

2–31. All bracelets are from the same company but are not good comparables.

If one is looking for provenance or background on some antique and period jewelry, often-overlooked sources of records are the Daughters of the American Revolution (DAR), Colonial Dames, and local historical societies. The local city chapters of all the above will generally have archival collections of photographs, letters, and/or documents that may prove vital for circa dating. All of the national offices of the above societies have archives that, upon the proper presentation of a request, are usually willing to share information.

With more and better circa dating on antique jewelry, a finer detailed description gives greater support to an estimate of value written by the appraiser. Proper identification is of such importance and impacts so highly on antique and period jewelry values that an entire section on identification techniques, clues, and motifs is presented in Chapter 3.

Using Price Guides

Price guides are not absolute and should not be considered as primary market values, but they *are* valuable tools for appraisers because they offer a form of stability. The professional valuer, using independently authored price guides for diamonds, colored gems, and antique jewelry has a basis for rational comparison of qualities and sizes versus costs.

There are numerous diamond and gemstone guides available by subscription in the United States and Europe. Knowing the strengths and weaknesses of the various publications helps appraisers use them more efficiently.

One guide with a section on antique jewelry is Ralph and Terry Kovel's *Antique and Collectibles Price List,* updated yearly. The list is valuable because prices stated represent actual retail asking prices. The Kovels insist that, while it may be possible for a buyer to negotiate a lower price, the strong point of the book is that none of the prices given is an estimate. If a range of prices is given, it is because they have found at least two of the identical items offered for sale at different places. A computer records various prices, they say, and prints the high and the low figures, but it *does not estimate.* In their guide a range can represent from two to eight sales of an identical item.

Martin Rapaport, author of *The Rapaport Diamond Report,* says that his price lists reflect the *maximum* price limit of the diamonds catalogued. "What it means," he explains, "is that a dealer will not sell for more than the amount in the guide." Most appraisers canvassed using the Rapaport Report say they are comfortable discounting his prices 20 to 25 percent in order to get a realistic estimate of the wholesaler-to-retailer diamond prices. Miami Master Gemologist Appraiser Joe Tenhagen believes that price guides only pinpoint a price range. He states, "The appraiser must check with two or three dealers in their own market that routinely handle whatever they are valuing. The price guides are built upon the New York markets and may not be an accurate reflection of one's regional market." He disputes the average appraiser's tendency to always discount Rapaport's list and warns that he can cite numerous instances where the price list's full price was asked for and received. "It all depends upon the geographical location of the appraiser and the strength of the supply and demand."

Anthony Valente of American Gem Market System, a telecommunication network in Moraga, California, cautions appraisers to look at the economic motive of the publisher of price lists as a gauge to how independent, or unbiased, they are. Valente says the price guides give an illusion of market volatility that is false. "We (AGMS telecommunication network) analyzed tens of millions of dollars in transactions over a six-year period, and we found that diamond prices were stable when enough data is counted versus one man's opinion or his brokerage activity." The strength of using a telecommunication network for appraisal prices, according to Valente, is that they analyze an enormous volume of sales that take place nationally. "The price lists are clearly for the purpose of selling against. It is a wish list, the prices the broker wishes to get. It is not reporting on actual transactions, and that is what one needs as a dependable basis for appraising."

For the appraiser who depends upon auction prices as a guide to valuation, only the hammer prices are considered viable estimates of value, and even those are conditional. No weight whatsoever is given to presale estimates because of the variables that can change the outcome of the sale, such as supply, demand, size of the audience, local economy, collector appeal, weather, and so on. Jeanenne Bell, author of *Answers to Questions About Old Jewelry,* writes that 70 percent of buyers at auctions are dealers. "Consequently," she says, "many of the prices reflect what dealers are paying and do not include a retail markup." A consensus of opinion by auction spokespeople from Sotheby's, Christie's, Butterfield's, and Skinner's currently estimates that 50 percent of buyers are dealers.

Emotion generated at an auction can distort the market value of an article, and two bidders locked in battle for an item can raise the price far beyond its true value. Conversely, it's impossible to know from auction catalogs if an item was sold or failed to meet its reserve. The reserve prices are generally 60–70 percent of the presale estimate, and knowing that figure will at least give an idea of what is considered bottom line for the piece by the owner and the auction house.

Antique jewelry price books are convenient for appraisers comparing styles but are limited in their use as actual value charts. They are generally outdated by the time they reach the market because it takes over a year for a publication to get from the author's hands to a bookstore. Also, most authors do not comment on condition or repairs, major elements in antique and period jewelry appraisals. Most jewelry price guides do not reflect the various strong regional markets for special types of jewelry. For instance, American Indian jewelry may sell for more money in New Mexico and Arizona than in North Carolina or Alabama. People are sensitive to the history and sentimental nuances of jewelry indigenous to their area, and so markets are created and kept alive.

Despite the inherent weaknesses in the information from books and guides, they remain the most useful tool the appraiser has. They are not a new commodity in jewelry appraisal, but they have taken on greater importance in the last few years. The prudent practitioner will remember they are subjective tools and not a substitute for diligent market research.

Publications

Suggested books listing antique and period jewelry prices:

The Official Price Guide to Antique Jewelry, Arthur Guy Kaplan; *Answers to Questions About Old Jewelry,* Jeanenne Bell; *Antique Jewelry with Prices,* Doris J. Snell; *Sotheby's International Price Guide; Fifty Years of Collectible Fashion Jewelry,* Lillian Baker; and *The Kovels' Antiques & Collectibles Price List,* Ralph and Terry Kovel.

Appraisers collecting auction catalogs will be interested to learn that, after a sale at Sotheby's, the catalog can be obtained for half price through the subscriptions department. At the end of the auction season, in July and August, remaining catalogues can be purchased for several dollars each; you can call by mid-July to reserve your catalogue. After the season, unsold catalogues are sent to Catalogues Unlimited for resale. Catalogues from all the major auction houses can be obtained for half price. Their address is P.O. Box 327, High Falls, NY 12440.

Judging Quality

Quality is expressed in two major ways: aesthetically and monetarily. As appraisers we want to learn to appreciate the aesthetic but be able to estimate and substantiate the monetary. All estimations of quality are relative; they vary from one appraiser to the other. Further, there cannot be an estimation of quality in the jeweler's art in relation to an absolute standard, since none exists. Consider this question: What is the highest point of excellence in jewelry quality? European jewelry? American jewelry? What about the delicacy, balance, and richness of style of Oriental jewelry? A standard of excellence means that other things are valued according to the closeness of their approach to its principles. If we chose one of the above as the standard of excellence, jewelry from folk artists and third world countries might fair badly by comparison. Therefore, we maintain that there cannot be an absolute standard of quality excellence for jewelry as a *single category,* but each type of item (ring, earrings, bracelet, and so on) must be judged in relation to others, identical or comparable. Some beginning appraisers who cannot judge jewelry quality have been known to rely upon a "gut feeling," or their own preferences. These are not yardsticks with which to measure quality.

Some gemologists, judging quality in antique and period jewelry, grant the piece merit upon the quality of gemstones only. While fine stones will increase the value of antique jewelry, the entire piece must be taken into consideration and judged by the standard of the time period that it represents. Diamonds and colored gemstones have not always been quality-graded the way they are today and cannot be judged by modern standards. More significant to the quality-grade may be an item's design, artistically crafted mounting, or embellishments. Judging quality on estate and antique goods requires studying hundreds of like items. When the research reaches into the thousands of comparable items, the appraiser can claim to be expert.

Recording Condition

Condition refers to how a piece of jewelry has been preserved. Several factors are considered in used jewelry: age, present risk, extent of previous repair(s).

An article may be in poor condition simply because it is *old,* and the natural deterioration of the materials has taken its toll. However, the poor condition may be due to physical or chemical abuse, or shoddy repair and disfigurement in the process. The appraiser should be able to recognize and describe damage that should have been repaired as well and describe damage that was not repaired.

Age is often the most difficult aspect of condition to describe and may be indicated by a wide range of danger signals. It can result from a natural weakening of the metals, as many materials tend to become weak and brittle over time. Brittleness is especially noticeable in organic jewelry materials such as tortoiseshell, shell cameos, ivory, and bone. Hairline splits and cracks are signs of the beginnings of problem areas. The joints of mountings and undersides of ring shanks should be examined closely for signs of repaired breaks, since these areas are vulnerable to stress.

Many metals and other jewelry materials are attacked by acids and environmental pollutants. If your client has trouble keeping the prongs on her ring for instance, she might be a daily swimmer in a chlorinated pool. Inquire about this possibility and caution her that chlorine will attack the gold and weaken the prongs, causing them to become brittle and easily broken.

Inspection of the condition of a piece of jewelry should take place under the same well-lighted conditions in which you inspect gemstones. Good illumination with cross-lighting and a loupe for magnification will help you detect small inconspicuous flaws, such as splits, tears, and breaks in metals and corrosion from environmental agents. Pay special attention to areas that have been cut for sizing, pin stems and catches, prongs, safety catches, and chain clasps. Gemstones and mounted coins should fit snugly in their holders or bezels without rattling around when moved. If they are loose, record it on the appraisal worksheet. For any possible future insurance claim, you want to have noted all the facts surrounding all items of jewelry being appraised.

Any damage found on the jewelry should be described as to its nature, location, and extent. Location can be mentioned using the terms *top, center,* or *bottom* of a specific design, or *top left* or *top right* of a specific central point. Also, a photograph showing the damage might be of value and provide a nice touch of professionalism.

Recording the extent of any defect may not be easy. While *dint, gouge, hole,* or *crack* is nomenclature that is in general usage and is easily interpreted, the general brittleness of an otherwise extraordinary shell or ivory piece may not be handily measured. Appraisers of fine art use the following adjectives to represent arbitrary degrees of defects: *negligible, slight, moderate, marked, extreme.* Slight refers to a defect more serious than negligible and less serious than moderate. It works well for fine and decorative art appraisers. Jewelry appraisers should seriously consider adopting this convenient nomenclature.

Publications

Catalogues of settings and mountings with prices include the following, which may be obtained from the individual companies, usually free of charge:

Albert Findings, Inc. 66 West 47th St., New York, NY 10036

Nassau Findings Corporation, 139 Fulton St., New York, NY 10038

AG Findings and Manufacturing Co., 55 NE First St., Miami, FL 33132

Ptak Bros., Inc., 2 West 46 St., New York, NY 10036

Other Market Information: Price Guides

Diamonds:

Rapaport Diamond Report
15 West 47th Street
New York, NY 10036

Diamond Market Monitor
66 Washington Road
Pittsburgh, PA 15228

The Guide
Gemworld International, Inc.
5 North Wabash
Chicago, IL 60602

Jean Francois Moyersoen
c/o Ubige, s.p.r.l.
Avenue Louise 221, Boite 11
B-1050 Bruxelles, Belgium

AGMS
1001 Country Club Drive
Moraga, CA 94556

Colored Gemstones:

The Guide
Gemworld International, Inc.
5 North Wabash
Chicago, IL 60602

American Gem Market System
GemWatch
1001 Country Club Drive
Moraga, CA 94556

Gem Connoisseur
Bennett-Walls, Inc.
P.O. Drawer 1
Rotan, TX 79546

Precious Gems Market Monitor
666 Washington Road
Pittsburgh, PA 15228

CHAPTER 3

MAKING CORRECT IDENTIFICATION

Proper identification is of primary concern to the appraiser. The three most important steps in producing an appraisal report and reaching a final estimation of value are: identification, description, and analysis. To possess authoritative knowledge concerning jewelry of any type, one must be informed about interrelated designs that have developed in various sections of the world, have some understanding of the commerce that has resulted in delocalizing the designs, and understand how jewelry is assimilated into the cultural and material history of many lands.

A broad view of the jewelry market rather than a focus on one small segment is essential. Unquestionably, one can identify jewelry by its generic name, that is, ring, bracelet, brooch, necklace, and so on; it is sometimes confusing to correctly identify antique, estate, and period jewelry. How can a value be researched for a component part of a necklace or brooch that has been removed from its original design and purpose if the origin and identity of that article is unknown? This is one reason professional appraisal skills embrace knowledge of styles, social customs, foreign cultures, and world history, along with gemstone and metal identification. The authoritative identification of jewelry objects is the most important of all value factors and the wellspring from which all other answers rise. In the final analysis, identification is the bedrock and basis of all value.

Figuring prominently in good identification is the objectivity of the observer. Every item can reveal itself in some manner, if the observer is aware of jewelry cycles, informed about social customs, a creative thinker, and a tenacious researcher.

Another attribute of the professional valuer is a discerning eye, the ability to recognize on sight the significant from the mundane. Deductive reasoning is a good trait to develop because early in the practitioner's career the appraiser should be able to quality-rank jewelry. How can one know what the best in the market is if one has never seen the best? A knowledge of quality measurements is important. To acquire this skill requires attention to U.S. and foreign manufacturing techniques, design innovations, trends, fads, and fashions. A good valuer must become a judge of quality because the value of a poor-quality article is considerably less than a similar one of high quality. To illustrate quality ranking in jewelry, consider the way an appraiser/gemologist quality-grades diamonds. It is accomplished in part by using standard color and clarity grading scales. The same scale procedure with slight alterations can be used to rank jewelry. Refer to the three items of enamelled jewelry in figures 3-1, 3-2, and 3-3, graded for the quality of the enamelling on each piece. A simple ranking system using a scale of *poor, good, excellent* helps the appraiser see what is the worst, mediocre, and best to provide a measure that can easily be interpreted and used to reach the final estimate of value.

Enamelling is a quality attribute that may easily be over- or undervalued. There are three kinds of enamelling: transparent, translucent, and opaque. Transparent enamelling reflects the colors used and gives a stained-glass effect; translucent admits partial light to the colored surface; opaque shows surface color only.

Enamelling has been popular in Europe and the Middle East for centuries, and jewelry with enamelled decoration will certainly find its way into the appraisal office. Sometimes the item has been enamelled not so much to enhance the design but to cover up poor craftsmanship, a ploy the appraiser should consider.

Enamelling has been used in jewelry as far back as the tenth century B.C., but as a fine art it began to

3–1. On a scale of enamel workmanship, this ring rates *poor.*

3–2. A *good* example of enamelling.

disappear in the seventeenth century. In the middle of the nineteenth century, the craft made a comeback as jewelry decoration. Because enamelling displays a gemlike quality, it is totally unlike anything else; when properly fired and finished, it can add value to an item. If treated well, it will suffer little deterioration and have a long life. When enamelling is poorly done and visually unattractive, value has to be deducted from the final estimated price. To aid the researcher in identifying and assessing enamel, a complete vocabulary of this technique may be found in the Glossary.

Learning to use a quality-ranking method is useful in the market data comparison approach. Reaching value conclusions via this course also requires market monitoring and hands-on research.

3–3. An *excellent* example of enamelling on this Art Nouveau brooch.

The most compelling reason for learning and using the ranking system is that by employing it one becomes not only connoisseur but expert.

Condition of the jewelry is important. When fine quality and pristine condition align, the value will be high, without question. It must be recognized, however, that there are numerous antique jewelry items where, although perfect condition is desirable, it is unobtainable due to age or abuse. Ancient seals, intaglios, and much of the jewelry from the Georgian period (circa 1714–1830) are such articles.

Using Symbols, Motifs, and Ornamental Styles as Identification Guides in Estate and Period Jewelry

One of the most expedient ways—and the key—to gaining expertise in identifying and valuing antique and period jewelry is by learning its history. The object is to research and compile historical knowledge of design, symbols, ornamentation, and decoration used in jewelry, textiles, and the decorative arts so that one may build a basis of comparison of jewelry in vogue today with the fashions of a thousand years ago. By tracking the use of similar motifs and jewelry styles over the centuries, the appraiser accumulates information that can be used to raise the level of expertise and ability to authenticate. Using the historical approach will impart an understanding of jewelry cycles, familiarity with the language of the past, the tools, and the attendant functions of the trade. Simply put, the exercise prepares the appraiser for general identification, witnessing, and estimating value.

Questions about style revivals, like Etruscan and Egyptian, can often be answered by the historical events of the period. For example, the treasures taken from Tutankhamen's tomb in the 1920s made Egyptian motifs the rage. The cycle repeated itself from 1976 through 1979 when the Egyptian government and Cairo museum held an exhibition of the King Tut treasures that traveled across America. This action was directly responsible for the revival of the Egyptian jewelry motif. The public was amazed at the fine goldwork and excited over the ancient designs and symbols.

When we speak of design, exactly what do we mean? There are several definitions of design. In simple terms, we may state that design is the orderly arrangement of lines and shapes. An orderly arrangement is necessary because without it there cannot be a real design. Behind order there is usually an idea to be expressed. There are two main divisions of design. *Pure design* is an arrangement of lines and shapes with the central idea of just looking beautiful without making a picture or telling a story. Representational design is pictorial, showing a picture of something real or imagined. One of the most graphic illustrations of representational design is found in the jewelry of the Art Nouveau period (1895–1915). Design applied to a flat surface is expressed in two dimensions: length and breadth. Repeated design is called *pattern.* A careful examination of the most intricate patterns show they are built upon a system of squares, diamonds, triangles, circles, hexagons, or other geometric forms. (All of this information is important, even crucial, to the accurate and full narrative description of a piece of jewelry on the appraisal report.) An item of jewelry that is carved, incised, embossed, or similarly decorated involves the third dimension, such as cameos and embossed mountings. As one becomes more familiar with this concept, it becomes clear that this type of jewelry belongs to the sphere of sculpture. Most jewelry belongs to the domain of pure design and is simply the idea of ornamentation and beautification. A brief survey later in this chapter speaks of the various influences on the development of design throughout history and notes certain characteristic features. With this information, one can place a design in its historic period, or revival period, for intelligent and satisfactory authentication.

The prevailing society and the theme of nature are the two most influential elements in tracking design logic. Nature influences are the topography, climate, flora, and fauna. The Egyptian lotus that grew so abundantly in the Nile is a design extensively copied by the artists of the day on all decoration including jewelry. The Persians borrowed from the palm, while the Greeks and Romans turned to the laurel and acanthus as the motifs of their art.

Other new forms are due to the effects of outside influences through trade or war. Egypt, Chaldea, and Assyria borrowed from each other, Greece borrowed from all of them, and Rome used all four.

The important social influences on design are religion, historical events, and trade. In many places and periods, religious laws have restricted artistic expression. The Muslims, forbidden by their faith to use living forms, human or animal, in their designs, turned to geometric motifs, especially the triangle.

Wars are historic influential events, and, as the invader entered a territory, new motifs in arts, fabrics, and jewelry began to appear. Napoleon's campaigns in Egypt in 1798 gave rise to the appearance of Egyptian motifs and subsequently to the first Egyptian jewelry revival.

Commerce and trade bring styles and motifs from one country to another. When the Dutch East India Company introduced oriental designs, such as the Chinese fret, into Europe, the rage known as chinoiserie (circa 1760) burst into bloom.

Familiarity with symbols dominant in certain countries, such as the scarab of Egypt, the cicada and dragon of China, and the laurel wreath of Greece, acts as another aid to identification.

The following breakdown in period motifs will allow the appraiser a more complete reference when considering the importance of motif, design, and ornamentation, and their relationship to identification.

Motifs and Designs Used in Various Cultures

Egyptian

This art begins with the Old Empire, about 4000 B.C. The motif most often associated with Egypt is the lotus. The flower appeared after the periodic overflow of the Nile and foretold abundant crops. Present in all its variations of flower, bud, leaf, and stalk, it is a sacred symbol of resurrection, fertility, and prosperity. If the appraiser is valuing *genuine ancient* Egyptian jewelry, more than just archeological interest must be studied. The designs of the goldsmith jeweler of the time not only reflected his skill and desire to beautify, but were important symbols that exercised a magic power on behalf of the wearer.

The following are typical Egyptian motifs: scarab or beetle, asp, cobra, papyrus, reeds, buds, date palm (symbol of reincarnation), chevron or zigzag (symbol of the Nile), scroll, fret, globe, disk, hawk, falcon, vulture, sphinx, and animal-headed human figures.

Babylonia, Assyria, Chaldea

These people had such interwoven histories and social structure that their designs must be viewed together. The people were warlike, cruel, and without deep religious conviction. The spirit of these people can be seen in their ornaments: the lion, man-headed winged bull, and the griffin. The chief motifs were: lotus, scroll, fret, guilloché, date palm, pine cone, lion, bull, and griffin.

Greece

Greek motifs, like the people themselves, were aesthetic, lofty, and intellectual. While many of the early ideas and designs were taken from other cultures, the Greeks amended and embellished designs with balance and grace. The Greeks aimed for a perfection of line and form. The ancient Greeks excelled in sculptural form; today their works serve as models in architecture and interior design. Greek artists and craftsmen were familiar with these motifs: lotus, anthemion, scroll, fret, dentil, guilloché, beads, reel-and-button, egg-and-dart, leaf-and-dart, acanthus, ram's head, and the human figure.

Rome

Roman motifs reflected every phase of life in ancient times as still seen in the well-preserved architectural remains. Clear detail about the character of these ambitious, practical, successful, and democratic people is evident. The motifs were not original but composites of all that preceded them. The Romans took the stiff motifs of Egypt, as well as the refined designs of Greece, and blended them into unrestrained, lavish ornamentals. More than two thousand years later, Roman motifs reappeared during the Italian Renaissance, in a more refined and delicate form. Some motifs that signal the Roman influence are: anthemion, scroll, fret, wave, guilloché, beads, borders, moldings, gadrooning, acanthus, human and animal masks, wreaths, swags, ribbons, garlands, fruit, oak, laurel, grotesques (hybrid human, animal, and floral forms), and warlike emblems such as the shield, spear, battle-ax, and eagle.

Byzantine

This culture had a fusion of western Asiatic and decadent Roman art. The most widely used motifs were: cross, crown, circle, lion, eagle, ox, angel, lamb, dove, peacock, fish, palm, grape vine, interlaced scrolls, braiding, basketwork, and pairs of birds or animals either facing each other or back to back and enclosed in roundels.

Late Persian

Persian motifs are strongly Asiatic and full of grace and symbolism. Lavish use of florals and animal forms, arabesques, and interlacing are seen. Most common of all is the pear-shaped motif, variously called the palm, pine cone, and ogee. The pear-shaped motif formed the basis for many European patterns, especially during the Renaissance. Other motifs: palm, tulip, rose, pomegranate, cypress tree, henna flower, trailing vines, animals, birds, human figures, and the paisley pattern.

Saracen

These nomadic people of the desert, sometimes called the Moors, conquered Spain in 710 A.D. and introduced Saracenic design into Europe. Restricted by Muslim law, Saracenic art did not depict human or animal figures in design. Using the triangle, the Saracens evolved a system of design that for richness of detail is unparalleled. The geometric motifs combined with foliage or vines are called *arabesques*. This art form lasted for one thousand years in Spain where it was known as Moorish. Motifs characteristic of this culture are: triangle, star, hexagon, shell, strapwork or interlacing, pea vine, arabesque, cartouche, and borders formed of monograms and inscriptions.

Gothic

The Gothic period, 1150–1500, produced the greatest art of the Middle Ages. The Gothic style originated in France and spread all over Europe, with its zenith of design and form development seen in the cathedrals of Germany, Italy, Flanders, Spain, France, and England. Gothic design is principally the circle and the evolved forms from the trefoil and quatrefoil. Gothic motifs are: trefoil, quatrefoil, rose, lily, clover, oak, grape, ivy, birds, animals, human figures, and grotesques. The grotesque was a fantasy shape that combined a human, animal, and plant form, such as a winged, human, female torso without arms terminating in foliage or in the hind legs of an animal. Much jewelry design of the Renaissance and later Art Nouveau period used grotesques.

Renaissance (Italian)

The Italian Renaissance covered three centuries and had three stages: early Renaissance (fifteenth century) in which art was expressed in simple forms and decorations were simple, high Renaissance (sixteenth century) when design became richly ornamental, and late Renaissance (seventeenth century) when designs and motifs evolved into the florid styles known as Baroque and Rococo. Since the Renaissance had distinct periods of development in various countries, we speak of the French, English, or German Renaissance. In Italy the early and high Renaissance saw use of the following motifs: wreaths, garlands of fruit and flowers tied with ribbons, urns, cornucopias, masks, cupids, grotesques, griffins, sphinx, dolphin, coat of arms, columns, and arches.

France

In the reign of Louis XIV (1643–1714), designs were semiclassical and grandiose. The "C" curve and curved acanthus foliage was used frequently on all types of jewelry. Designs were of symmetrical arrangement with favorite motifs being cartouches, masks, grotesques, strapwork, birds, feathers, shells, ogee, lattice, Chinese motifs, monograms, war trophies, caryatids (female figures used as the main structure of a piece of jewelry, for instance, a ring shank), ribbons, and lace. During the reign of Louis XV (1715–1774), use of the Chinese motif, or chinoiserie style, was popular as were graceful bows and arrows, hearts, flaming torches, baskets of flowers, bowknots, shepherds' crooks and hats, and landscapes with groups of figures carved on cameos. The Louis XV style evolved into the Rococo period with the "S" and "C" scrolls, stripes, lattice design, shell, ribbons, garlands, feathers, lace, flowers, birds and small animals, lovers in gardens, pastoral settings, cupids, and doves as prominent themes.

Cupids have been ambiguous figures in old jewelry, so it is good for the appraiser to bear in mind that this symbol was used on jewelry to adorn the living, and also as a memorial piece to pay tribute to the deceased. A close inspection of the item and an inscription may help separate the jewelry's role. However, if the inscription is of no real value or the piece is without inscription, look at the colors of the item, if colors are present. Bright colors indicated life and love; dark colors were used for somber occasions.

The period of Louis XVI (1774–1792) was a time of classical design with symmetrical arrangement. Favorite motifs of the day were guilloché, egg-and-dart, fret, running wave, beads, fluting, fillet, acanthus, flower sprays, garlands, musical instruments, and garden tools. Since Marie Antoinette enjoyed both music and role playing at gardening, both the musical instrument and musical notes adorned textiles and jewelry, along with garden designs.

The era of Napoleon I was called the Empire period (1804–1814) and ushered in a return of the heavy and formalized styling of classical motif. The favorite themes were: sphinx, griffin, lion, goddess of victory, victory wreaths of oak or laurel, palm branch, anthemion, star, coronet, medallions, fleur-de-lis, bee, letter N, eagles, and war trophies.

England

After one hundred years maturing in Italy, the Renaissance reached England. In England it became a mixture of Gothic and Classic, with old Moorish designs and new local motifs, such as the Tudor rose, rambling floral and leaf forms, and heraldic emblems. Over time the Gothic, Spanish, and Flemish motifs blended into a typically English style in which the flora depicted became more realistic. By then, the reigning monarchs were William and Mary, followed by Queen Anne.

The Georgian period (1714–1830) brought a demand for enamelled jewelry, diamonds, and items of delicate Classical motifs. As this was also an era in which archeological achievements were being heralded, designs of Pompeii were revived. Other popular motifs were arabesques, grotesques, husk flowers, rosettes, beads, drapes, and triple plumes.

Germany

German craftsmen showed little feeling for classical design throughout the entire Renaissance period. Although they were superior craftsmen, they were content with florid, detailed styles of decidedly Baroque attributes, and almost no new national designs emerged.

Why is all the foregoing important to the appraiser developing expertise in period and modern jewelry identification? Because when one is conscious of how the spread of design has influenced craftsmen on all continents, the following relationships are more easily understood:

1. The work of designers or jewelry craftsmen of different periods contain basic elements in common, and some similar patterns and symbols will be produced no matter how dissimilar the native designs. (The message: don't make swift judgments or attribute a work to a specific maker or period without adequate research.)
2. Sometimes the foreign element in design is disregarded. (The message: knowing design origin helps the appraiser select and defend the origin of the jewelry.)

To be efficient as an appraiser requires a keen watch on the work of foreign, domestic, arts and crafts, and commercial jewelry makers. Just like designers of long ago, today's jewelry craftsman looks to the past to select styles and motifs that can be imitated and expanded upon. The appraiser must be able to tell the difference between the genuine article and the imitation.

If distinguishing between the revival items or *old* reproductions of earlier periods is a puzzle, consider that all artists and jewelry craftsmen *in all periods* were surrounded by the artistic idioms of their times. They had no knowledge of future styles and they could only copy the preceding styles. Therefore, the appraiser looking at an old reproduction must know the peculiarities of specific designers (Fabergé, Castellani, Giuliano, and

Table 3-1. Periods of Style, Design, Motif

Historical Style	Original Design	Motif Revival		
Egyptian	3000 B.C. to 300 B.C.	1780–1820	1870–1910	1922–30
Greece	700 B.C. to 150 B.C.			
Rome	700 B.C. to 400 A.D.	1790–1830	1850–1870	
Greco-Roman				
Renaissance	1350 A.D. to 1550	1780–1820	1850–1900	
Rococo	1600 A.D. to 1750	1870–1900		
Moorish	800 A.D. to 1650	1850–1870	1910–1930	
Celtic	650 A.D. to 1100	1880–1910		
Gothic	1150 A.D. to 1500	1810–1830	1855–1880	
Etruscan	800 B.C. to 300 B.C.	1850–1890	1925–1935	

Compiled by Tom R. Paradise, reprinted with permission.

so on) and their styles in a given period. For example, the cameo faces carved in the Greek and Roman era will exhibit certain characteristics: Roman nose, curling locks of hair, godlike attributes. In twentieth century copies of such works, faces take on the character of twentieth century personalities since the artist works in an entirely different era and milieu, and this difference will show at once when compared with an authentic Roman or Greek period cameo. To use this technique as a means of identification requires study of genuine examples until the variations become apparent. The astute appraiser may develop a sense of perspective in time, and find that a work cannot have, for example, neoclassical elements and belong to the Art Deco period. It *is* either neoclassical or from the hand of someone careless about details. Learn to read the jewelry, make it tell you what it is!

Tom R. Paradise, GG, FGA, of Atlanta, Georgia, has made an in-depth study of motifs, designs, and styles and correlated their use to various jewelry periods (see table 3-1). Paradise uses this subject to lecture at seminars and classes on antique jewelry. Some illustrated terms are presented to help the jewelry appraiser become better acquainted with this important element.

Decorative Motifs

The appraiser should be familiar with the following terms used to describe various patterns, borders, and motifs. All of these terms are illustrated in figures 3-4 through 3-29 and the descriptions below have been compilated with the assistance of Tom R. Paradise.

- *Acanthus:* The acanthus is a plant common in southern Europe. Greek, Roman, and Byzantine cultures all used this classical motif. It appears in Gothic, neoclassical, and Rococo decorations.
- *Arabesque:* Ornamentation named for motif believed to be of Arabian origin. Of the several types of arabesque, the most often seen is a style of interwoven figures, fruit, and flowers. The design is found in many French items from the seventeenth-century Renaissance until the first quarter of the eighteenth century.
- *Banderol:* Spiraling ribbon decoration.
- *Boss:* Decoration like a rosette used at an intersection where four elements, usually lines, meet. A Gothic style of ornament.
- *Celtic Interlace:* Originated in runic or Celtic basketwork.
- *Chevron:* Zigzag pattern.
- *Chimera:* A winged or goat-headed lion used to symbolize strength and agility.

3–4. Acanthus.

3–5. Arabesque.

3–8. Celtic Interlace.

3–9. Chevron.

3–6. Banderol.

3–7. Boss.

3–10. Chimera.

3–11. Dragon.

3–15. Fretwork.

3–12. Egg-and-Dart.

3–16. Paterated and Guilloché.

3–13. Festoon.

3–14. Foliate.

- *Dragon:* The European dragon looks like a cross between a snake and a crocodile and is said to be related to the hundred-headed hydra that guarded the Garden of the Hesperides. The European dragon thrived upon a diet of virgins and was usually slain by a hero or god. Dragons were a sign of warfare to the Romans. In Christian art they represent sin and Satan. The Chinese dragon (illustrated in fig. 3-11, from the Ming period) connotes good instead of evil, symbolizing the Emperor. It is the Spirit of the Waters. Dragons with five claws appear on items for imperial use; dragons with four claws or less are for members of the imperial household and officials. The Japanese dragon has three claws.
- *Egg-and-Dart:* A variation of the Bead-and-Reel design. This dates from Greek times and was used in neoclassical period jewelry. It consists of oval (egg-shaped) and dart-shaped figures.
- *Festoon:* A swag of garland of sagging shape.

3–17. Grotesque.

3–19. Laurel and Acanthus.

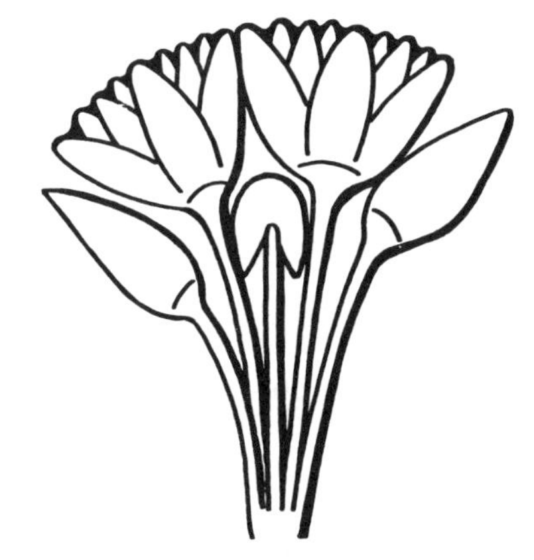

3–20. Lotus.

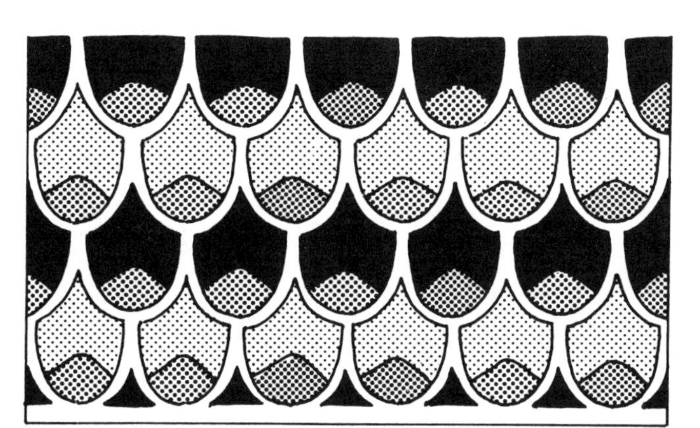

3–18. Imbricate.

- *Foliate:* Any leaf or plant design, such as pine, laurel, acanthus, holly, or palm.
- *Fretwork:* A repeated geometric design as in the Greek key pattern. Such geometric patterns were also called *meanders.* The fretwork motif developed independently in ancient China is found on early bronze items.
- *Paterated* and *Guilloché*: Paterated is a repeated circular design. Guilloché is a repeated double serpentine design.
- *Grotesque:* A fantastic shape that is usually a combination of human, animal, and plant form. Commonly used in ancient Pompeii and revised in the Renaissance and again in the Art Nouveau period.
- *Imbricate:* Fish-scale pattern.
- *Laurel:* The laurel motif appears frequently on jewelry of the eighteenth century. Illustrated in figure 3-19 is a combined laurel and acanthus pattern.
- *Lotus:* This was the most common motif in Egyptian art and jewelry.

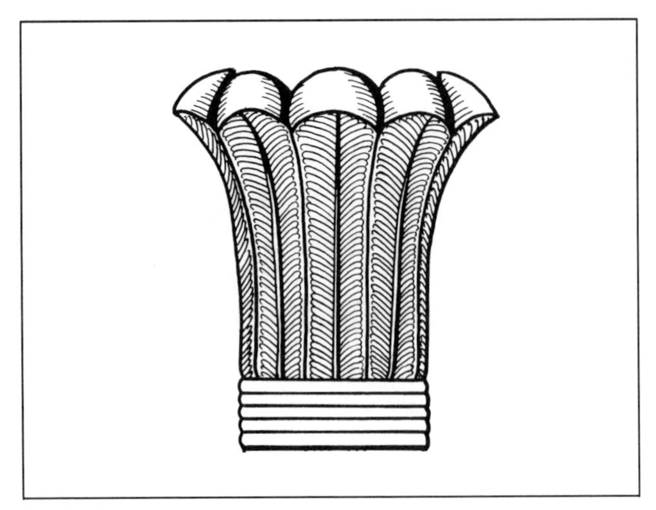

3–21. Palm.

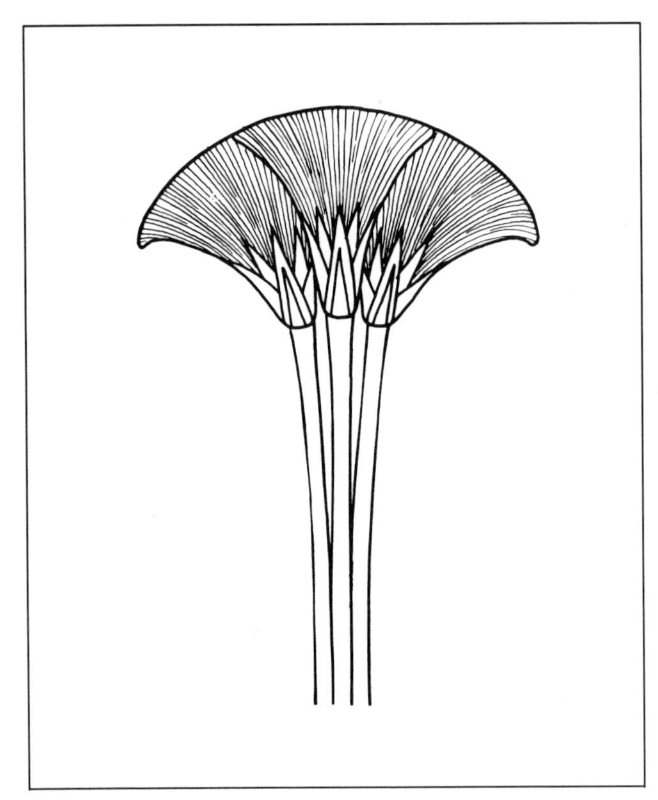

3–22. Papyrus.

3–23. Rinceau.

- *Palm:* Frequently used as decoration during classical periods and a basic design of the Egyptian period (including revivals). The palm is a symbol of victory and peace.
- *Papyrus:* This is one of the most important motifs in Egyptian art. The papyrus is a leafless stem with a triangular cross section and flowers in a large umbel at the top.
- *Rinceau:* Repeated serpentine foliate patterns, sometimes seen as grapevines.
- *Rococo (Baroque):* Swags, garlands, and scrollwork were often associated with Louis XV and the era of the 1750s.
- *Shell:* Marine shell designs were used as motifs of decoration in the Rococo period. The use continues today. The scallop shell, shown in figure 3-25, is most commonly used.
- *Tesserae:* Individual mosaic pieces. One example would be the fragments in a piece of micromosaic jewelry.
- *Trellis:* A gridlike design.
- *Vitruvian Scroll:* A scroll with several convoluted undulations.
- *Volute:* Spiral decoration.

3–24. Rococo (Baroque).

3–25. Shell.

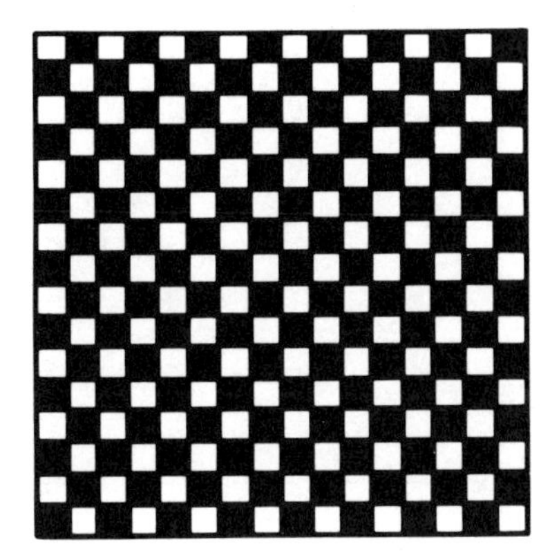

3–26. Tesserae.

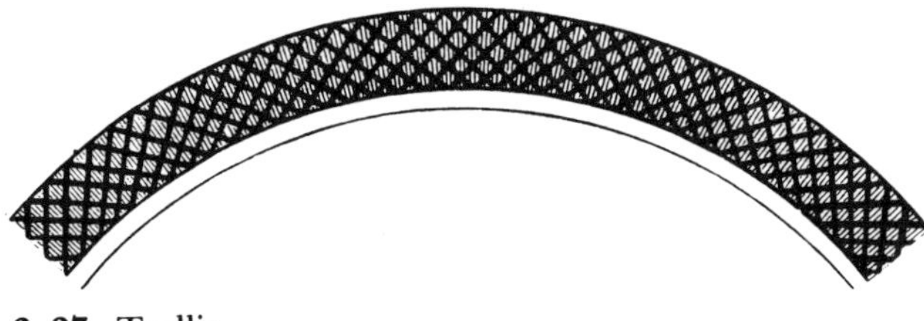

3–27. Trellis.

3–28. Vitruvian Scroll.

3–29. Volute.

Using Color as a Clue to Jewelry Dating

The use of color in gemstones and enamelled jewelry parallels the story of color in architecture and the decorative arts. Its expression in ancient times was vigorous. It became gaudy during the Renaissance and turned solemn during the Victorian era. It was during the Victorian era that jet, made from a particular kind of coal, came into vogue when Queen Victoria, after the death of her consort Albert, declared an extended period of mourning. Other gloomy-colored material used for jewelry was bog oak and gutta percha. Following the Victorian era was a period of soft, almost delicate, color with richly colored gold and silver metals and colored enamels—Art Nouveau. This period was short lived, as the fashion for muted color changed when the bawdy Art Deco of shockingly bright hues and collections of strongly contrasting colors followed. Art Deco was an interesting reflection of the devil-may-care mood of the people. As a style, Art Deco returned in the 1950s and revived once more in the 1970s, but with less gusto.

Knowing about color helps appraisers because the understanding of how colors dominated certain periods puts a finger on the pulse of the time. It also provides a tenet of observation that leads to more precise jewelry dating and greater overall appraisal expertise. Knowing what gemstones were predominent at different times helps the appraiser since colors too run in cycles of popularity and supply. GIA Librarian Dona M. Dirlam has compiled a chronological list of gemstone discovery, table 3-2 (see pp. 38–39), that points up the alliance of gemstones and colors with various periods.

The obvious use of color by ancient people was in the belief of its usefulness in working magic. Every color had its significance. The usual intention was to avert the evil eye or to secure good luck. This was especially helpful when used in combination with gemstones in which the color of the stones could work their magic according to material and hue. To be set in jewelry reinforced the spell because of its innate amuletic properties.

The ancient Egyptians, Greeks, Tunisians and Turks used color brilliantly and adhered to simple hues with symbolic meanings. There was a strong attraction to triad arrangements: red, yellow, blue; red, blue, white; deep red, medium yellow, blue; deep blue, light blue, white; blue, ochre, black. The vast quantity of mosaics made as far back as 4000 B.C. have been found with patterned textures in color. There are records of vivid color schemes used in both textiles and jewelry.

Table 3-2. Chronology of Gems*

Gem Material	Discovery	Source	Reference
AMBER	Ancient	Baltic Sea	Ogden p. 90
BERYL	Ancient - 3rd c. B.C.	USSR, Turkey	Ogden p. 93
Aquamarine		Iran	Sinkankas p. 9
Emerald	Ancient - 2nd millennium	Egypt	Ogden p. 92
		India	Sinkankas p. 9
	1500s	Colombia	Sinkankas p. 21
CHRYSOBERYL	1830	USSR	Webster p. 130
Alexandrite	1932, 1975	Brazil	G&G 1988 p. 16
Cat's-eye	Ancient	Sri Lanka	Ogden p. 93
		Brazil	Webster p. 131
Chrysoberyl	Late 1600s	Sri Lanka	Bauer p. 303
	1805	Brazil	Bauer p. 303
CORUNDUM	3rd c. B.C.	Sri Lanka	Ball p. 274
Ruby	1597	Burma	Ogden p. 95
	1800s	Thailand	Webster p. 84, Bauer p. 286
Sapphire	3rd c. B.C.	Sri Lanka	Ogden p. 94
	1800s	Thailand	Bauer p. 286
	1865	Montana, USA	Webster p. 87, Bauer p. 289
	1881	Kashmir	Webster p. 85, Bauer p. 288
	1870	Australia	Webster p. 86
	1960s	China	G&G 1988 p. 155
Star Sapphire	Ancient		Ogden p. 94
	400 B.C.	India	Webster p. 17
DIAMOND	600 A.D., 14th c.	Borneo	Webster p. 20
			G&G 1988 p. 67
	1725 (1714)	Brazil	Webster p. 21
	1829	USSR	Webster p. 45
	1849	USA	Webster p. 23
	1851	Australia	Webster p. 23
	1866	South Africa	Webster p. 32
	1908	Namibia	Webster p. 36
GARNET	3rd c. B.C.	India	Ogden p. 90
		Sri Lanka	Ball p. 275
		Egypt	
IOLITE	Ancient	Sri Lanka	Ogden p. 94
		India	Webster p. 331
		Burma	
JADE	18th c.	Burma	Webster p. 255
Jadeite	Early people	Central America	
Nephrite	Ancient	Asia	Ogden p. 99
			Ball p. 213
	1000 B.C.	Central Asia	Webster p. 255
	Early people	New Zealand	Webster p. 258
	1850	USSR	Webster p. 258
JET	Ancient	Whitby, England	Ogden p. 90
LAPIS LAZULI	Ancient - 3rd millennium	Afghanistan	Ogden pp. 40, 114
	Early people	Chile	Webster p. 252
MALACHITE	Ancient - 7th millennium	Egypt	Ogden pp. 101, 103, 114
OPAL	Ancient - Roman	India	Ogden p. 104,
		Eastern Europe	Webster 233
	(1849) 1872	Australia	Webster p. 234
	1200	Mexico	Webster p. 234
	1906	USA	Webster p. 239
	1970	Brazil	Webster p. 239

Table 3-2. Chronology of Gems (*continued*)

Gem Material	Discovery	Source	Reference
PEARLS	Ancient	Waterways, worldwide	Kunz 1907
PERIDOT	Ancient-Pre-Dynastic	St. John's Island (Zabargad), Egypt	Ogden pp. 102, 104
QUARTZ/CHALCEDONY			
Amethyst	Ancient	USSR	Ogden p. 105
		India	
		Brazil	G&G 1988 p. 214
Carnelian Agate	Ancient-Pre-Dynastic Egypt	Egypt	Ogden p. 91
		India	
		Europe	
Citrine	Ancient-Hellenistic Roman	Western Europe	Ogden p. 106
		USSR	
		Sri Lanka	
		Spain	
Rock Crystal	Ancient-Pre-Dynastic Egypt	India	Ogden p. 106
		Egypt	
		Europe	
Rose	Ancient	Europe	Ogden p. 106
		USSR	
Smoky	Ancient-3250 B.C.	Europe	Ogden p. 107
		USSR	
		Sri Lanka	
		India	
SPINEL	Ancient-late Roman	Sri Lanka	Ogden p. 111
		Afghanistan	Ball pp. 64, 274
SPODUMENE	1879	North Carolina	Webster p. 166
TOPAZ [confused with peridot and citrine]	Ancient-Roman	India	Ogden p. 111
		Sri Lanka	Ball p. 166
	1737	Europe	Webster p. 144
TOURMALINE	Ancient-Roman	India	Ogden p. 111
		Europe	Bauer p. 364
		Iran	
	Late 1600s	Sri Lanka	Webster p. 146
		USSR	Webster p. 151
		Brazil	Webster p. 151
	1820	USA	Bauer p. 371
		Mozambique	Webster p. 151
	1970	Kenya	Webster p. 151
TURQUOISE	Ancient-5000 B.C.	Iran	Ogden p. 113
		Europe	Ball p. 213
		Egypt	Webster p. 244
	Ancient	Mexico	Webster p. 241
		USA	Webster p. 248
ZIRCON [confused with hessonite]	Ancient	Sri Lanka	Ogden p. 114
		Thailand	Ball p. 136ts
		Burma	
		Australia	
ZOISITE			
Tanzanite	1967	Tanzania	Webster p. 370

*Compiled by Dona M. Dirlam, GIA.

Table References: Ball S.H. 1950, *A Roman book on precious stones,* Los Angeles, CA: Gemological Institute of America; Bauer M., Spencer L.J. Translator 1969, *Precious stones,* Rutland, VT: Charles E. Tuttle Co.; *Gems & Gemology:* Various Issues; Kunz G.F., Stevenson C.H. 1908, *The book of the pearl,* New York: Century Co.; Ogden J. 1982, *Jewellery of the ancient world,* New York: Rizzoli; Sinkankas J. 1981, *Emerald and other beryls,* Radnor, PA: Chilton & Co.; Webster R., Anderson B.W. 1983, *Gems, their sources, descriptions, and identification,* London: Butterworths.

In the Orient, red, yellow, and gold have always been the dominant colors. The Greeks favorite colors were blue, yellow, red, green, purple, black, and white. The Greeks were responsible for the earliest known attempts at color shading.

The Dark Ages were certainly dark in regard to jewelry. Not until the Italian Renaissance was the art of color enamelling revived on jewelry.

In Latin American countries, notably those that abound in precious gems, professional rank has traditionally been noted by the colored gemstones worn. A physician wears an emerald ring, an engineer wears a sapphire, and an attorney wears a ruby. The green tourmaline is the special stone for a professor, while the man who runs or works in a commercial enterprise wears a pink tourmaline. The precious topaz is the special stone for a dentist.

Colored birthstones correlate to months and zodiacal signs, and in Thailand nine-stone colored gems good luck rings are made and sold daily. The nine stones used and believed to bring the wearer good fortune are diamond, ruby, sapphire, garnet, black onyx, zircon, pearl, cats-eye, and emerald.

Among American consumers, rubies, sapphires, and emeralds are still the most popular colored gemstones. However, demand for certain stones also follows the choice of colors in fashion and costume. In the marketplace today, professional color forecasters are predicting the popularity cycle of different colors for many years into the future. If their predictions are correct, and they have been up until the present time, colors will be a neon-bright eclectic mix and all colors will be of intense hue. Purple, which used to be worn only by kings, will be a favorite of the consumer.

Cut of Stones

Occasionally, the cut of set gemstones in a piece of jewelry may help determine the age of the item. The square outline of the old mine-cut and the rose-cut were popular during the Georgian era. The old mine-cut began to fade away about 1910. The old mine-cut, baguette, briolette, step cut, and navette were all used during the Victorian period. The rose-cut appeared in the sixteenth century and was revived toward the end of the nineteenth century. The old European-cut stone was used throughout the Art Deco period and began to disappear from modern cutting about 1920. It is difficult to put exact dates on the discontinuation because long after gem cutters were rounding up stones, others were still using the old styles. Sometime between 1910 and 1920, most cutters were bruting octahedra to get the circular girdles. Also, introduction of the saw early in the century did away with grinding away the tip of the octahedron because it became possible to saw either through the center or near center to produce the proportions we associate today with the 58-facet full-cut modern round brilliant diamond.

Supplemental Identification Information for Antique Jewelry

Antique jewelry is that over one hundred years old. Many pieces the appraiser handles will not be true antiques but rather heirlooms or period jewelry.

Antique jewelry collecting is growing in popularity because many see it as a link to the past. Some items, however, are not purchased from dealers but are part of a family heritage and passed from generation to generation. Thus, it is not unusual to be asked for a jewelry appraisal by a forty-year-old client who insists the item is an antique because it was owned by her great-grandmother. If the jewelry's style or maker's mark can tell you immediately it is not over one hundred years old, there is a diplomatic way to get this across to the client in a logical and convincing way. General antiques experts Ralph and Terry Kovel use the following simple formula for quick computation of the *possible* age of an item: divide the client's age in half and add twenty-five years for each past generation of owners. By this system the client's jewelry in the above example would be computed as twenty years, with twenty-five years added for grandmother, plus another twenty-five years added for great-grandmother. The total of seventy years is a possible age for the jewelry item limited to outside conditions, such as the item being purchased as an *antique* by the grandmother. In any event, this is an interesting exercise for your client. Used in a gentle and persuasive manner, it can help the client see the difference between an object being antique and merely old. Naturally, any computations as above are subject to additional research by the appraiser if antiquity is an issue to value.

Researching antique jewelry offers a fascinating historical perspective of where we have been as craftsmen, designers, interpreters of beauty, and provides us with a mirror reflection of our ancestors' lives. Like supporting players in a stage production, antique and period jewelry will reveal their role in the drama of their time provided the audience—in this case the appraiser—is alert and pays attention.

One's familiarity with the costuming of certain periods, hairstyles, necklines, hemlines, sleeve lengths, use of gloves or bonnets, as well as social preference for materials such as brocade or muslin, can be the pivotal factors in a correct identification and estimation of value.

It is interesting to note that many outside influences have altered the course of fashion and design and weigh heavily upon the use and wear of jewelry: social changes, wars, religion, and cultural and economic shifts.

Before the eighteenth century, jewelry was worn because of its symbolic importance and its visual impact upon the masses about the great wealth and importance of its owner. Gems and jewelry played important roles, which defined one's position on the social scale. Following the French Revolution and under the new democratic social order, anyone could wear any item of jewelry he or she pleased regardless of social position. As a result of the new social order, jewelry became mostly ornamental to enhance the fashions of the day.

A library of information about period and antique jewelry and the influences surrounding its wearing is invaluable to the appraiser. The following chronological period overview is a brief look at the major motifs, styles, designs, costumes, manufacturing methods, gems, and market influences that set trends and helped make jewelry history.

The Georgian Period 1698–1830

The illustration, figure 3-30, pictures costuming during the period around 1660; the illustrated figure 3-31 shows what the fashionable woman was wearing in 1810.

Gemstones in general use: The Georgian period was known as the age of the faceted stone, and diamonds were popular. The early cut styles most often used were the Mazarin or table-cut, rose-cut, briolette, and pendeloque. The table-cut diamond was a cushion-shaped stone with seventeen facets both above and below the girdle, and a culet. A brilliant-cut with fifty-eight facets was introduced by a Venetian cutter named Peruzzi at the end of the seventeenth century.

Other popular gems were: almandite garnets, emeralds, topaz, agate, amber set in silver over red foil, colorless quartz, and river pearls. Glass (paste) was introduced to the French court by Georges-Frédéric Strass in 1780. An Englishman, George Ravenscroft, had produced flint glass in 1676 and lead crystal in 1681.

Popular jewelry materials: gold in 10, 18, and 22 karat; silver, silver over 18K gold (most popular way to

3–30. Mid-seventeenth-century dress. (*Illustration by Elizabeth Hutchinson*)

3–31. A stylish lady, circa 1810. (*Illustration by Elizabeth Hutchinson*)

set diamonds), cut steel, and pinchbeck (a material of 83 percent copper and 17 percent zinc, a popular gold imitation).

Manufacturing methods: handcrafting, sand casting. Settings had closed backs, and most gemstones were enhanced with a backing of silver or copper foil. As advances in gem setting and the nature of light on gemstones was understood, the settings were opened for more brilliance and better dispersion. This did not occur, however, until late in the eighteenth century.

Motifs: bows, hearts, flowers, ribbons, birds, stars, wheat stalks, floral sprays, and garlands.

What the jewelry was like: By turn, cluster and pearl drop earrings were fashionable. In 1720 a French edict from the court of Louis XV forbade the wearing of diamonds, pearls, and precious stones among French women and ordered jewelers to export what they had in stock. The edict was in force for six months. Meanwhile, in England, earrings were very popular, and in 1774 a lady's magazine advised women that it was in the best of fashion to wear small drop earrings. Garnet jewelry was also becoming more simply set at this time and more geometric in design.

The *rivière* (rivers of stones) diamond necklace was a great favorite of fashionable ladies. Double-row rivières date from the beginning of the period, and rivières with triple rows of diamonds from the end of the nineteenth century.

A favorite adornment was the diamond aigrette. This was a jeweled feather hat, brooch, or hair ornament shaped like an egret plume. It was set solidly with small gemstones, usually diamonds, and had top stiff projecting wires that trembled with any movement of the head or body. Sometimes the projections were finely coiled springs. Jewelry expert Joyce Jonas of New York University lectures that if the appraiser sees an aigrette where the flowers or stalks are soldered and cannot move, the piece is devalued by one-half. If the piece also suffers from a poor and careless soldering job, the piece may receive as much as 90 percent devaluation. In fact, Jonas warns, the only value left in the item might be the rose-cut diamonds!

Market Influences:

- Indian diamonds were replaced by diamonds discovered in Brazil in 1725.

English monarchs:

George I	1698–1727
George II	1727–1760
George III	1760–1820
George IV	1820–1830
William IV	1830–1837

French royalty:

Baroque period, Louis XIV	1643–1715
Chinoiserie trend, Louis XV	1715–1774
Fashion simplicity, Louis XVI	1774–1792
Classic period, Louis XVII	1793–1795

Napoleon seized power in 1799 and ruled France from 1804–1814

Louis XVIII	1814–1824
Charles X	1824–1830

The Victorian Period (1837–1901)

The Victorian Period ranges from Victoria's accession to the throne until her death in 1901. This period is easier to understand if segmented into three parts: early Victorian, the romantic period from 1837 to 1860; mid- or high Victorian from 1860 to 1885; and late Victorian from 1885 to 1901.

A characteristic of Victorian-era jewelry is the lack of a maker's mark or hallmark on the jewelry. The British government did not require jewelers to use markings on the products during most of the nineteenth century. Much of the gold jewelry produced was 18K until about 1854 when the use of 9, 12, and 15 karat gold was made legal in order to meet foreign competition. Having that positive date is a help in circa dating.

Early Victorian Period (1837–1860)

The illustrations, figures 3-32, 3-33, represent costumes of the 1840s. Clothes at that time were demure and covered all of the body. No open necks meant no necklaces, and since bonnets or hair covered the ears, earrings were not worn. Extremely large brooches were in fashion and they were worn both with the low décolletage that prevailed in evening dresses, or at the neck in a high-necked day dress. Jewelry historian Margaret Flower reveals that the hand was the center of interest with rings being fashionable as well as large bracelets with pendants. This style of bracelet made the hand of the wearer look dainty and feminine.

During the 1850s, clothing became rich and elegant in material and form. Custom and style called for a center parting in the hair which made an excellent frame for wearing a diadem. Earrings reappeared and necklaces were worn with evening dress. Bracelets alone or in pairs were in fashion.

Gemstones in general use: The early Victorian period generated a prolific use of the following gemstones: diamonds (rose-cut and brilliants), amethyst, pink and golden topaz, turquoise, chalcedony, coral, garnet, ruby, cameos carved from hardstone, shell, lava, and coral. During the 1820–1840 period, carved mala-

3–32. Costuming in the early nineteenth century. (*Illustration by Elizabeth Hutchinson*)

3–33. Ears were covered with bonnets or hair, so earrings went out of fashion. (*Illustration by Elizabeth Hutchinson*)

chite was fashionable in jewelry use, with parures set with malachite cameos or in thin carved plaques. Seed pearls were popular for young ladies. Hair jewelry was very popular and flourished.

Popular jewelry materials: gold from 18K to 22K, and tricolor gold, silver, rolled gold, gold electroplate, and pinchbeck.

Manufacturing methods: Hand manufacturing was still predominant; however, the Industrial Revolution was beginning to make inroads into jewelry production. Jewelry historian Joan Evans writes of the French jeweler Beltête who in 1852, suffering from rheumatism in his fingers, invented a mechanical process for cutting out and stamping settings.

Motifs: coiled snake, angels, love knots, arrows, bows of ribbons, Celtic, crosses, daggers, daisies, doves, flowers, Gothic, hands, ivy, Moorish, scrollwork, stars, strapwork, vines, Classical Greek and Roman designs, eye miniatures, clover, triangles, butterflies, bars of music from a hymn, heavy festoons or garlands, swags (fig. 3-34, p. 44), and Algerian knots (fig. 3-35, p. 44).

What the jewelry was like: Fortunato Pio Castellani of Rome was producing classically designed Roman jewelry in archaeological motifs from gold. The items were decorated with filigree. Jewelry with filigree, cannetille, colored gold, or small pearls were representative of the articles bought by the middle class.

The motif of a bird defending its nest against a snake came into fashion on a brooch or pendant about 1835 and was in demand for twenty years.

The Algerian campaign (1840–1860) made a pendant, brooch, earrings, or hairpins fashionable when designed with the elaborate knots and tassels that were a part of the Algerian dress. The Moorish style followed as a fad.

There was a trend to brooches and bracelets shaped like boughs of wood in 1848. In 1850 bracelets in enamelled *plaid* designs with a diamond buckle were the rage, also brooches and lockets with receptacles for locks of hair, sentimental jewelry; and velvet ribbon around the throat with ends crossed in front and fastened with a brooch.

During the period 1838–1858, bracelets with heart shaped pendants, brooches with hanging pendants, hoop earrings, and trembling ornaments as brooches, and for the hair, were popular.

3–34. Festoon-style necklace with cameos, circa 1870. (*Photograph courtesy of Tom R. Paradise, T. R. Paradise & Co.*)

Market Influences:

- Queen Victoria greatly influenced the jewelry fashions. She was a diamond lover (diamonds in rose-cut and brilliant-cut), and the new rich of Britain's industrial society followed her example.
- Etruscan tombs opened in 1836 at Caere, and the Thomas Cook tours of the pyramids set the Egyptian revival style of the 1850s.
- Tiffany's opened in New York in 1837.
- Large scale jewelry manufacturing began in the United States in the 1840s.
- Napoleon III and Eugenie (the Second Empire) returned to favor in 1852, but their rule collapsed in 1870.
- The Metropolitan Opera House (New York) opened in 1883. The first tier of seating was called the Diamond Horseshoe because of the brilliance of the diamond jewelry on the women patrons. It is reported that the splendor of their jewelry outshone the house lights!

Mid-Victorian Period (1860–1885)

The costume (fig. 3-36) shows what was worn during the years 1875–1880, an era of dashing styles. Strong color was encouraged in both clothing and jewelry. In the early years of the period, necklaces, brooches, and bracelets were elaborate and heavy. Earrings were long and dangling and lockets were in favor once again as women began to wear open collars.

By the 1870s earrings had become even longer with use of pendants or fringe to accent the length. Decorated gold surfaces were carried to the extreme in ornamentation. Faceted gemstones were themselves set with a motif in the center of the gem. A backlash against excessive ornament in the 1880s made a plain strand of amber beads the only acceptable jewelry. This was short

3–35. Algerian-knot motif Victorian brooch. (*Photograph courtesy of Mark Moeller, GG, CG*)

3–36. Costume of the mid-Victorian period. (*Illustration by Elizabeth Hutchinson*)

lived and gave way to a penchant for plain gold bangle bracelets, small button earrings, and chains.

Gemstones in general use: amethyst, carbuncles (garnets in cabochon cut), coral, crystal, enamels, diamonds, emeralds, onyx, opal, pearls, rubies, tiger-claws set in gold, turquoise, sapphires, black glass, amber, bog oak, ivory, jet, and tortoiseshell. During this period some ghastly fads appeared that were, mercifully, short lived: live beetles with jeweled backs were tethered to a tiny gold chain on a pin-and-stem and then were allowed to wander across the bosom as a kind of living brooch. Genuine stuffed hummingbird heads dangled from feminine ears!

Popular jewelry materials: gold, both low karat and imitation; silver, plain and oxidized; steel.

Manufacturing methods: handwork, some die-stamping, limited casting. The Etruscan motif jewelry was decorated with gold grains and fine gold wire.

Motifs: acorns, amphorae, anchors, monograms, snakes, crosses, hearts, crosses, bees, bells, birds, daisies, flowers, geometrical, insects, masks, mottoes, shells, sphinx, stars, swans, and memorials.

What the jewelry was like: The best designed jewelry of the period was being produced by Italy's Fortunato Pio Castellani, England's Carlo Giuliano, and Boucheron in France. Castellani produced jewelry with filigree, granulation, and enamelling as his specialty. His hallmark is the interlaced back-to-back "C." Giuliano, previously employed by Castellani, opened his own shop in England in 1875. His hallmark, "CG", or "C and AG", will be found on his granulated gold and mosaic work. Frédéric Boucheron, a leading French jeweler whose firm was founded in Paris in 1858, was noted as a designer of luxury jewelry.

Jewelry of the period included the following popular pieces: high Spanish tortoiseshell combs worn with a mantilla, strings of pearls entwined in the hair, tiaras, aigrettes, cabochon stones with motifs such as diamond or pearl stars or flowers set in the stone rings, long crosses, large heart pendants, gold necklaces with pendant urns, acorns, or masks, jeweled and enamelled necklaces with small drops, heavy gold chokers, thin gold chains, medallion pendants. In cameos the female figure was decorated with jewels. This type of ornamented cameo is called *habillés.* Brooches inscribed with mottoes, painted enamel miniatures in filigree frame brooches, and large presentation brooches were popular. Flowers or birds in jewels and brooches with a Celtic influence were in vogue.

Market Influences:

- Gold strike in California in 1849
- Australian gold rush in 1851
- Gold stamping law in United States in 1854
- Commodore Perry's arrival in Japan in 1854 inspired Japanese-motif jewelry
- Death of Prince Albert in 1861
- The Civil War in the United States, 1861–1865
- Diamond discovery in South Africa in 1867
- First transcontinental railroad completed in 1869
- Introduction of electricity as source of power and light in 1870

Late Victorian Period (1885–1901)

The woman pictured in figure 3-37 wears a costume that was favored for evening wear about 1895. She wears a low neckline to show off pearls or a necklace, and a pleated and embroidered bodice that precluded the need for a brooch. This period emphasized curves of the figure. There was an abundance of lace on collars and in bows for daytime wear, fulfilling the need for ornamentation at the neck. The jewelry was small and lighter than in previous periods, because clothing material was lighter weight (silks, taffeta) and produced a more delicate look. Rings and bracelets were of a narrow width, earrings were tiny studs, if worn at all. Hair was swept up on the head with wide-brimmed

3-37. Costume of the late Victorian era. (*Illustration by Elizabeth Hutchinson*)

hats for day and silk flowers or jeweled ornament decoration for evening. Because of lightweight fabrics, heavy brooches were swept out on the tide of fashion, while small pins that were worn scattered on the bodice came in. For evening wear the diamond pins were preferred, and they were often placed in the hair. It also became the fashion to stick pins in the folds of lace or bows of ribbon and on bonnets and hats.

Gemstones in general use: amethyst, aquamarine, chrysoprase, diamonds, emeralds, peridot, rubies, moonstones, chrysoberyl, opals, sapphires, and turquoise.

Popular jewelry materials: gold, silver, oxidized silver, platinum, horn, and rolled gold.

Manufacturing methods: As the nineteenth century wore on, handcrafting jewelry, although never completely forsàken, gave way to machine-produced jewelry in mass production. Demand from a growing middle class of consumers turned the wheels of commerce. A jewelry historian has written that the age divided itself into three categories: designer, craftsman, and salesman. The art of handmaking jewelry was suspended.

Motifs: crescent, Etruscan designs, stars, bows, knots, feathers, flowers, quatrefoil, trefoil, clover, cross-over, double hearts with crown or knot, doves, Egyptian motifs, horseshoe, cats, moons, owls, swallow, turban, rose, oak leaves, grape clusters, edelweiss, fox masks, and hunting or sporting motifs.

What the jewelry was like: bar brooches, choker-type necklaces, brooches with invisible settings, narrow bangles. Safety pins joined by chain to a brooch were in demand. Solitaire diamond rings set in gold or silver gained a market niche in 1895. Stomachers (a large brooch that is worn in the center of the midriff and directly under the bosom) were favorite evening ornaments. Stud earrings were desirable and class rings were beginning to grow in popularity.

Market influences:

- Invention of the automobile in 1895
- Spanish-American war in 1898
- Death of Queen Victoria in 1901
- Edward VII and Alexandra ascend the throne of England, reigning from 1901 to 1910.

Edwardian Period (1901–1914)

The Edwardian period costume illustrated in figure 3-38 is a reflection of an elegant period marked by gentility and gracious good manners.

Queen Alexandra, wife of England's Edward VII, was the beautiful style-setter. The style she chose for daywear was one of simplicity. The gowns were made in lightweight materials, frequently in white or some subtle pastel hue, and had slender waists to accent the popular "S" shape. Other garment accents were lace, embroidery, fringe, and fur.

3–38. Costume of the Edwardian period. (*Illustration by Elizabeth Hutchinson*)

Alexandra was fond of flowers and often wore real ones on her dress; therefore, jewelers produced gold flowers with enamelled petals in a wide variety of flower-designed brooches. Crosses as well as long gem-set chains were appropriate day jewelry.

The high collar dresses provided good backdrops for the popular brooch, while evening gowns with off-the-shoulder styling were perfect frames for the multistrand dog collars favored by Alexandra. Rumors were that the queen wore the dog-collar style to hide a small scar on her throat. For whatever reason, fashion or vanity, a trend for the choker-type necklace developed.

As a backlash to the heavy and solid diamond jewelry sprays of the nineteenth century, diamond jewelry was made as fragile looking as possible. The metal platinum best accomplished this by using a light and airy type technique known as a knife-edge setting. This style made diamonds look as though they were suspended in midair because only the thinnest line of metal was seen.

Gemstones in general use: diamonds, peridot, amethyst, pearls, emerald, sapphire, ruby, garnets, opals, moonstones, mother of pearl, and all forms of enamelling.

Popular jewelry materials: gold in multicolors, platinum, silver, oxidized silver, rolled gold, celluloid.

Manufacturing methods: machine-made, mass-produced, and handcrafted.

Motifs: horseshoe, wishbone, garlands of flowers, swags, leaves, shamrocks, hearts, bows, scrolls, crescent moon and stars, single hunting dog or hunting scenes, insects, man in the moon, moon and owl, and sporting scenes such as race horses.

What the jewelry was like: Diamond aigrettes or diamond feathers in the hair were popular. Long ropes of pearls, which had been in relative obscurity until the end of the nineteenth century, were combined with diamonds for evening wear. The pearl sautoir, a long rope of pearls ending in a tassel, first popularized after the French Revolution, was revived. Pendants in the form of round medallions with diamond borders and delicate sawed and pierced center sections were testimony to the skill of the era's craftsmen. Demantoid garnets set into reptile design pins were popular. Peridot and kunzite were in vogue because Alexandra loved the color pink and green was Edward's favorite. Gilded and enamelled insects were produced as brooches and in a variety of pendant designs.

Market influences:

- Edward VII and Alexandra, 1901–1910
- Wright brothers' first air flight in 1903
- First moving picture in 1903
- Great San Francisco earthquake and fire in 1906
- George Frederick Kunz, employed by Tiffany & Co. in 1879, studied American freshwater pearls and helped bring them prominently to the market. He discovered a variety of pink spodumene in California that was named kunzite in recognition of his efforts in the mineralogical field.
- 1900 Paris Exposition
- France's involvement in Morocco, the Turkish wars, and the Balkan crisis created an oriental influence in French jewelry styles.
- Cartier opened a New York branch in 1909.

Art Nouveau Period (1895–1915)

The costume of the period is illustrated in figure 3-39. The era after 1870 was a grand time of peace and prosperity in the world, combined with low taxes, luxury living, and a general feeling of well-being. The social season required a large wardrobe: morning dresses, afternoon gowns, evening gowns for dinners, balls, opera, and the theater. Since the opera and theater gained prominence in England, men and women were required to have cloaks, mantles, shawls, and jackets for evenings out. Of course, this called for many types of appropriate jewelry, such as brooches, parures, matched sets of necklaces, brooches, pins, earrings, bracelets, and rings. Each outfit had to have its own jewelry.

Until 1913 the neck was covered in the daytime by a high tight collar of net or lace. Occasionally this would be set off by wearing a dog-collar necklace. Hair was styled in a relaxed upsweep. Dress fabrics were of fine texture with an immense amount of embroidered handwork trim. The hobble skirt made a brief appearance in England in 1910 but never caught on in America.

Gemstones in general use: synthetic rubies, emerald triplets, crystal, chrysoberyl, carnelian, lapis lazuli, malachite, demantoid garnet, pearls, tourmaline, small diamonds, ivory, horn, moonstones, opals, amber, and enamels.

Popular jewelry materials: platinum, white gold, silver, yellow gold.

Manufacturing methods: The jewelry of the Art Nouveau period shows the beginning of modern jewelry design. Pieces are interesting because of their machine work or very fine handcrafting. Concurrent with this period was the Arts and Crafts Movement, a rejection of machinery and a glorification of the designer/craftsman. In its purest form it never caught on and did not produce many pieces. Stamped pieces were common.

3–39. Costume of the Art Nouveau period. (*Illustration by Elizabeth Hutchinson*)

Motifs: graceful insects such as dragonflies, plants, flowers with coiling tendrils, snakes, lizards, maidens with flowing hair. Abstracts such as sunbursts and crescents had wavy curves, such as the whiplash curve. The Oriental influence was strong, particularly in the French items.

What the jewelry was like: This was also a period of Egyptian revival jewelry as well as Japanese-influenced designs. The collars, cuffs, ribbons, and so on, were detachable and many were held in place by "lace pins," delicate open work or filigree small pins. A small 10K gold pin called a negligee collar pin, resembling a small bar pin, came into vogue about 1901. They were also called handy pins and worn at the collar or neckline. The dog collar made popular by Alexandra often had as many as nineteen rows of tiny pearls kept parallel by diamond openwork bars. The richest women wore pearl sautoirs, 45-inch-long necklaces of pearls falling to the waist and pinned to the bust with a jeweled watch. Long chains held muffs for the fashionable woman to warm her hands. Muff chains were an Art Nouveau specialty.

Market influences: Several artisans exerted great influence in the jewelry market. Among them were Charles Ashbee and Harry Wilson of England, René Lalique of France, Louis Masriera of Spain, Josef Hoffman from Austria, and Peter Carl Fabergé of Russia. All were mainstream designers in the development of Art Nouveau jewelry styles.

Art Deco Period (1920–1930)

Art Deco is all about geometric designs, abstract patterns, and wild eclectic combinations of colors in floral designs. Although there was no precise day when Art Deco began, many point to the arrival in Paris of the Ballet Russe in 1910 as a major influence on the period. The vivid color combinations of sets and costumes in the production of *Scheherazade* were immediately adopted by artisans in other fields of the decorative arts, from interior design to jewelry. The Art Deco name is derived from the *Exposition des Arts Décoratifs et Industriels Modernes* held in Paris in 1925.

Since it was an era of change, and freedom of expression was permissible, women cut their hair, shortened their hemlines (fig 3-40), bound their busts for a thin boyish look, and took up smoking. The short hair, which made aigrettes and jeweled combs obsolete, sparked a revival of earrings, and screw-back fittings were invented for those who did not want to pierce their ears. In the 1930s, when hair was worn longer, earclips were stylish. Women wore close-fitting cloche hats; designers made little brooches especially for the cloches. Strapless and backless dresses were part of the evening-wear trend; these called for long dangling pendants and sautoirs of pearls, the flapper look.

With the growing acceptance of women smoking, the cigarette case was born. Up until 1914 it was a misdemeanor for women to smoke in public! With the rise of moving pictures, the stars of the silent screen became the new dictators of fashion in both garments and jewelry wear.

Gemstones in general use: diamonds, emeralds, rubies, sapphires, black onyx, crystal, ivory, jade, coral, mother-of-pearl, synthetic stones, carved colored gemstones. New cuts and shapes of gemstones emerged with emphasis on geometry. Baguettes, emerald-cuts, triangles, and shield-cuts were popular.

Popular jewelry materials: Platinum replaced yellow gold as the favored metal; white gold, silver, plastic, chrome, and marcasite.

Motifs: geometric designs and abstract patterns, cubism, Egyptian, African, Oriental, and American Indian designs and motifs. Floral-designed jewelry with carved Indian rubies, sapphires, and emeralds gave a fruit-salad look to items that designers and the public enthusiastically accepted. The Oriental influence was felt in carved coral, jade, and other gemstone flowers, dragon, and Fu-dog motifs. Enamel enhancements were popular for all jewelry items. The 1922 opening of King Tutankhamen's tomb sparked another Egyp-

3–40. The Art Deco period was the era of the flapper. *(Illustration by Elizabeth Hutchinson)*

tian revival in design, and Egyptian motifs taken from border designs, pictorial images, and hieroglyphs were chic.

What the jewelry was like: There were huge brooches and brooches worn in pairs called clips that were placed on necklines, belts, hats, and shoes. Jabot pins were smart. The new design cut from Paris, the French-cut square diamond, was set in flexible box bracelets (the straightline bracelet). Several new gemstone cuts had arrived from Europe and all were popular: the kite-cut, hexagon, square, half-emerald, lozenge, and pentagon. Platinum link bracelets in all widths were greatly favored by the American woman. Necklaces were long thin chains or strands of lapiz lazuli, pearls, coral, jade, or agate beads. The diamond was the dominant gemstone and large diamond solitaires were fashionable. Unusual combinations of materials characterize Art Deco jewelry: coral and diamonds, black onyx and diamonds, turquoise, sapphires and quartz crystal. Sporting themes were popular in brooches, especially diamond pavé creations like sailfish and marlins. In the late 1920s, machine-made flexible box bracelets set with diamonds were matched up with ladies wristwatches with newly miniaturized movements and the diamond-encrusted cocktail watch was born.

Leading designers: A number of great designers contributed to the period. Some of the most well known were Jean Fouquet, Gerald Sandoz, Paul Brandt, Louis Cartier, and Frédéric Boucheron. The great houses of Mauboussin and Van Cleef & Arpels kept the period lively with innovative designs. In the United States the firms of Lalique; Tiffany; Winston; Marcus and Co.; Black, Starr and Frost; Spaulding & Co.; C.D. Peacock; Shreve, Crump and Low; and Galt & Bros. were leading jewelers of the period.

Market influences: The United States of the 1920s was a consumer society. Not only the affluent but ordinary men and women bought jewelry, not because of need but for the sheer pleasure of buying. The two most powerful instruments of culture in the period were the radio and the movies.

A financial disaster and social crisis occurred with the crash of the stock market in 1929. This was followed by the Great Depression of the 1930s. The public gave up their jewels as payment on loans. Banks found themselves with a flood of jewels on their hands, some from jewelers in distress. The Depression produced a drastic price drop in natural pearls, which fell to one-tenth of their previous market value.

The vitality of the period was never regained even after the economy improved. The geometrical style was outdated and suddenly associated with unhappy times.

The reproduction: In the 1980s a revival of the Art Deco period brought its attendant reproductions. Some of the better-made pieces crafted in Portugal, Spain, and Japan can be difficult to detect. Particular attention should be paid to the following: method of manufacture, finish, gemstone quality, and overall condition.

The finer pieces of genuine Art Deco jewelry were handmade and some have hand engraving and/or millgraining. The gemstones were usually channel-set or bead-set in the original jewels. A genuine Art Deco era item will be well finished; in many instances the reverse side or underside is as well articulated as the top. An old piece, even from the 1920s, should show some natural wear; even gemstones may show some surface scratching or light abrasion. Also, colored gemstones and diamonds in the genuine Art Deco pieces should not display obvious enhancements such as heat treatments, irradiation, or lasering.

Retro Period (1940–1950)

The costume of the Retro period (fig. 3-41), looks astonishingly like the 1980s' fashions. The dress, except for the long gloves, looks modern.

3-41. The costume of the Retro period looks amazingly modern, pointing up fashion's penchant for cycles. (*Illustration by Elizabeth Hutchinson*)

The well-dressed woman of the early 1940s helped the war effort by working in factories and industry. She adopted slacks and culottes as appropriate wear. For those working with machinery, about the only item of jewelry that could be worn without worry was an ankle bracelet. The snood was worn as a hair cover and is enjoying a fashion revival in 1989.

Dior created fashion news in 1947 when he showed the new feminine look with longer, fuller skirts, clinched waist, and rounded shoulders. The jewelry used to accessorize the new style was huge: massive rings, earrings, and brooches. The little black evening cocktail dress was a fashion statement along with a tailored suit. Both required large gold accessories with big colored gemstones or diamonds for proper accent. Cultured pearl necklaces were very popular.

Gemstones in general use: aquamarines, golden beryl, citrine, peridot, and tourmaline. Most of the stones were of extra large size to accommodate the scale of the jewelry. During the war years diamonds, sapphires, and rubies remained popular and were plentiful in the United States.

Popular jewelry materials: Until 1941 platinum was the most widely used metal for jewelry. However, when the United States entered the war, the government forbade the creation of platinum jewelry. Platinum was a strategic metal needed in the war effort. Jewelers turned to white gold, but it was not well accepted by the consumers who considered it more costume than fine jewelry material. Then the United States restricted the use of gold in 1943.

Palladium became the reluctant next choice for jewelry, but manufacturers did not like using the material because it was so difficult to work. However, the restrictions on both gold and platinum were rescinded in September 1944, much to the delight of jewelers and designers.

Motifs: flowers, scrolls, ballerinas, American flags, animals, shells, birds, hearts, cupids, American Indian themes, and baskets of flowers.

What the jewelry was like: gold chains and choker sets with large citrines, topaz, or aquamarine stones.

Dresses had bracelet-length sleeves, so jangling charm bracelets, especially collections of silver hearts, were designed to fill a fashion need. Also, ropes of cultured pearls, snake-link chains, clusters of colored gemstone brooches in flower motifs, collections of bangle bracelets worn together, large single-stone rings, and plaque pendants were all considered chic wear.

The pompadour hairdo for women created the perfect frame for the clip earrings that were worn by day and the long dangling earrings that were reserved for night wear. The majority of earrings were either clip or screw-backs, because pierced ears were out of style and considered a barbaric custom—a holdover from another era.

Market influences:

- World War II; the United States entered the war in 1941.
- Women entered the work force in large numbers, many for the first time working outside the home. This produced a cultural change in the United States.

Modern (1950–1980)

The costume of the era (fig. 3-42) is a commentary on the trend of the last three decades in the United States toward the youthful look. The postwar period saw the arrival of short skirts and a flood of jewelry for a luxury-hungry American market. In the mid-1950s the chemise, a kind of unbelted plain garment that required plenty of jewelry accessories, was fashion's darling. The sack dress was set off with long ropes of beads, bib of pearls, bracelets, and earlobe-to-shoulder dangling earrings. Pierced ears returned to fashion for women, and to men who dared.

Gemstones in general use: in the late 1950s rubies, sapphires, emeralds, peridots, and diamonds, in the 1960s peridot, colored sapphires, spinels, tourmalines (especially rubellite), emeralds, rubies, amethyst, and yellow and blue sapphires used in combination were popular. The late 1960s saw two newcomers to the market, tanzanite and tsavorite.

Popular jewelry materials: yellow gold, sterling silver, new materials such as niobium, titanium, and tantalum. Old materials like wood, ivory, bone, leather, and plastics were reintroduced. The U.S. government deregulated the price of gold in August 1971 and the price climbed from $35 an ounce to an all-time peak of

3–42. The style of clothing in the last two decades reflects the cultural accent on youth.

$850 in 1981. The result was lighterweight jewelry mountings with heavy sculptural looks. Tiffany's pioneered a trend for sterling silver jewelry; however, during the precious metals price rise, silver prices also skyrocketed. All manufacturing processes were utilized: handcrafting, cast, and die-striking. Also, many items were combinations of the manufacturing methods.

Motifs: animals, snowflakes, stars, flowers, sea urchins, bouquets of flowers, pine cones, peace signs, yin-yang symbols, male/female symbols, wings, rockets, symbols of space travel, geometric shapes, Egyptian motifs, and American Indian themes.

What the jewelry was like: The jewelry of the 1960s was influenced by the youth counterculture movement with items in natural materials like wood, bone, and ivory. Ornate East Indian filigree pieces and American Indian silver and turquoise jewelry swept the country. Puzzle rings and ankle bracelets were popular. Rings were worn on every finger and two of them on the index finger. Men wore more jewelry than at anytime in the last five decades; chains, ropes of beads, heavy bracelets, and earrings. Cuffs were popular for the arm worn with two bracelets. The necklaces were pulled into a knot instead of using a clasp; medallions were stylish.

The 1970s ushered in a return to a more classic and simpler jewelry style. Elsa Peretti designed her famous bean, teardrop, and heart-motif pendants. Coin jewelry and zodiac jewelry were popular.

In the late 1970s jewelry artists such as Picasso and Calder began to exhibit sculptural jewelry. The stars of soap operas such as *Dallas* and *Dynasty* boosted the jewelry industry with their display of jewels, much as the Hollywood stars did in the 1930s and 1940s. Princess Diana of the British royal family made the sapphire and diamond wedding ring popular, and certain rock star heroes created fads for pins or pendants.

Market influences:

- Korean War
- Hippie movement
- Space flights
- Civil Rights
- Vietnam War
- Inflation and recession

CHAPTER 4

ESTIMATING VALUES

Finger Rings

From ancient times until today, a circle known as a finger ring remains the most popular item for personal adornment. It has been used as a symbol of power, slavery, love, and remembrance. In some periods of history, both hands were covered with rings worn under and over gloves, and in every possible shape and style. There are hints that Neanderthals may have worn rings because rings of rushes, weeds, bones, and stones were in use about 3000 B.C., well before the era of metal technology.

Who invented the first rings as we know them? History gives parallel credit to both the Phoenicians and Egyptians. To get answers to the questions as to why and how the first rings were worn, however, would require the researcher to wade through a torrent of legend, fiction, conjecture, and superstition. The written records are often confusing and contradictory. Most experts on ancient rings will agree, however, that men—not women—were the first and only ones to wear finger rings for a very long time. References to men wearing rings in single and multiple styles, together with a list of privileges conveyed by wearing such ornaments, are found in texts from Pliny's *Naturall Historie* to George Kunz's *Rings for the Finger.* The Gauls and the Britons issued codes on ring-wearing by men. In the sixteenth century there were lists issued assigning the wearing of rings to the following: the thumb for doctors, the index finger for merchants, the middle finger for fools, the annular (third) finger for students, the auricular (small) finger for lovers.

Gold and iron rings were worn in ancient Persia, but gold rings were regarded as an emblem of fidelity in a wedding ceremony, while any type of ring was felt to be an emblem of eternity. Rings from the Bronze Age found in archaeological digs show some to be of bronze, some a type of wire, and some of gold.

One of the most ancient rings ever found came from an excavation in Egypt. The hieroglyphic inscription is for Cheops, the pharaoh of the famous pyramid tomb.

One of the most interesting records found with an excavated ring, as reported by James McCarthy in *Rings Through the Ages,* is a jeweler's sales guarantee of 429 B.C. Translated it reads:

> "As concerns the gold ring set with an emerald, we guarantee that in twenty years the emerald will not fall out of the gold ring. If the emerald should fall out of the gold ring before the end of twenty years, Bel-ah-iddina, Bel-shumu, and Hatin (the royal jewelers) shall pay unto Bel-nadin-shumu an indemnity of ten mana of silver."

There seems to be very little new in the world of commerce!

The custom of wearing rings went from the Greeks and Etruscans to the Romans. Greek rings were made of various materials including gold, silver, iron, ivory, and amber. The Etruscans are well known for their magnificent gold work, and Napoleon III was said to have treasured an authentic Etruscan gold ring. Some Etruscan rings removed from tombs are in the knot motif, others as coiled serpents. Romans wore iron rings, at first as a mark of honor; later, all freemen were permitted to wear one. A gold ring, at first restricted to be worn only by the emperor, was granted to senators and finally to soldiers. During the reign of Tiberius (A.D. 14–37), it is reported that the wearing of the gold ring was only granted to those whose fathers and grandfathers had property valued at 400,000 sestertia. In time, the wearing of gold rings became an accepted fact, for women as well, and attention was turned to ring styles. What followed was a kind of Roman orgy in the prodigious display of rings worn several to a finger, many engraved and gem-set. Other customs and fads connected with the Roman penchant for wearing rings included key rings (actually keys to homes and vaults), memorial rings, and natal rings. Since the birthday was one of the most solemn events in the life of a person, the birthstone ring was important to the individual.

During the Middle Ages a period of staleness in ring design set in that did not lift until the fourteenth century. However, to restrain the masses from too much display or ornament, Edward III issued an edict that all persons in England under the rank of knighthood or having less than two hundred pounds in land or chattels could not wear rings or other jewelry.

Rings as expressions of love, betrothal, and marriage reach back as far as ancient Egypt. The ring most commonly examined, identified, and valued by appraisers is the engagement and wedding ring. It is appropriate, therefore, to have some historical knowledge of this form of adornment in order to amplify the appraiser's ability to identify and increase expertise.

The Egyptians believed that a vein ran from the third finger of the left hand directly to the heart. Since a ring was an object of emotion, it should be placed upon that finger. The Romans exchanged rings at the conclusion of a business contract, and this included marriage contracts. A plain iron hoop was most commonly given as a symbol of the cycle of life and eternity and was a public pledge of betrothal. Interestingly, betrothal and nuptial rings have not always been worn on the third finger of the left hand as is the custom today. Just when and where the switch to the left hand came is not clear, but when the giving of a ring became part of the Christian marriage ceremony in the ninth century, the ring was placed on the bride's right hand. As can be seen in wedding portraits from the thirteenth to the sixteenth centuries, this practice continued in France until the fifteenth century. Historians tell us it was well into the middle of the eighteenth century before Roman Catholics began to use the left hand for nuptial rings. Today, in some European countries, the right hand still receives the nuptial ring; Norway is an example. In some countries in the Middle East, it remains common practice for the wedding ring to be placed upon the first finger.

By the fifteenth century, the diamond had become the symbol of conjugal faithfulness and part of the ritual of weddings. The use of a diamond ring for betrothals appeared towards the end of the fifteenth century. A letter written in 1477 to Archduke Maximilian before his engagement to Mary of Burgundy says, "At the betrothal, your Grace must have a ring set with a diamond and also a gold ring." The Sforza marriage ring, pictured in figure 4-1, is an Early Renaissance ring copied from an ancient manuscript in the Vatican. It is significant because it documents an important marriage in the fifteenth century. The ring is set with a natural rough diamond that is half hidden in the ring setting. The upper half of the octahedral crystal rises from the bezel and reflects light from all four sides of the exposed upper portion of the stone. Instead of feeling constrained by the heavy closed mountings, the late medieval goldsmiths used imagination to design and enhance the gems set in rings. By the end of the fifteenth century a breakthrough in diamond-cutting techniques offered a new shape in diamond rings. The diamond had a table-top cut, the pyramidal point rubbed flat with diamond dust. This was the first step towards what we know today as modern cutting technique. The table-top cut became an important feature of sixteenth-century diamonds, and can be used as a guide to dating jewelry.

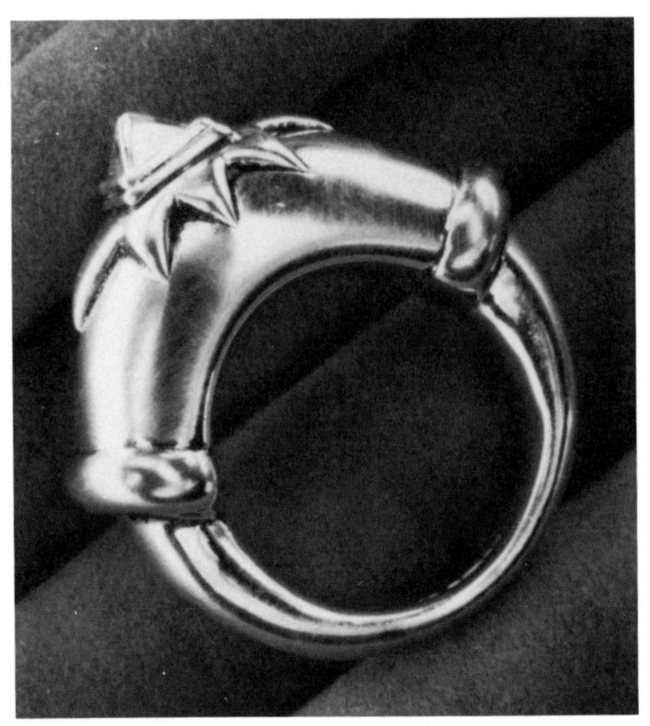

4–1. The Sforza marriage ring, a copy of the late medieval-style ring found in an old manuscript in the Vatican. It documented an important, fifteenth-century marriage in the Sforza family. A poem found in the manuscript emphasized the important role of the diamond in sealing the union:

Two torches in one ring of burning fire
Two wills, two hearts, two passions are bonded in marriage by a diamond.

(Information and photograph courtesy of Diamond Information Center).

The gimmal (from Latin, meaning twins) ring in figure 4-2 consists of two or three interlocking bands,

4–2. An antique type of wedding ring still used by some Irish couples.

4–3. A modern-day version of a flower bouquet motif.

usually with a heart motif on the center band and one hand on each of the outer bands. The bands are joined by a pivot so that when brought together the hands clasp and enclose the heart and look like a single ring. Gimmal rings are sometimes called Una Fede rings and were popular as wedding rings in the sixteenth century. The style remains popular today as a friendship ring.

Betrothal rings fell out of fashion for a period and lost their status as romantic necessities before they were revived and renamed *engagement rings.* Revival and resurgence of the ring was primarily due to the historic South African diamond discovery in 1867. For the first time in history, supply, demand, and price were aligned and the stone became universally favored for use in engagement rings.

The variety of engagement ring styles used throughout the decades is interesting and can be used for approximate jewelry dating. The marquise diamond cluster ring was popular in the time of Louis XVI and retained its fashion statement for over 150 years. The shape of the mounting is a long pointed oval running in the same direction as the finger, but instead of being set with a single marquise-cut diamond, the mounting contains a cluster of smaller diamonds. Cluster diamond rings in the round flower-bouquet motif, like the modern-day copy in figure 4-3, were said to be very popular in Martha Washington's day. They were set with rose-cut diamonds that usually surrounded a larger center rose-cut diamond.

Of special interest in motif research is the emphasis on mountings with scalloped edges and galleries. According to writer James McCarthy, these were known as *cookie* rings because of the delicacy of design. A modern-day interpretation is illustrated in figure 4-4.

The hoop ring (usually a plain gold band when it is a wedding ring), is a simple bold band embellished with

4–4. The *cookie* ring of Martha Washington's time had rose-cut diamonds in rows around a center rose-cut diamond. The mounting border was scalloped; sometimes there was a small well for perfume hinged on the underside of the ring.

a single diamond set deep in the mounting. This type of ring is recorded throughout Colonial and early American history. The style is still used today and, when diamonds are set partially or all around the band, it is called an eternity ring.

In the eighteenth century Queen Charlotte of England received so many pieces of diamond-set jewelry from King George III and the Indian rulers that she was nicknamed the Queen of Diamonds. The Queen wore a *keeper,* or diamond hoop ring, like the one in figure 4-5 on page 56, to protect her wedding ring.

A forerunner of the diamond solitaire was the Princess ring, a mounting with three to five diamonds set in a straight row across the top (fig. 4–6). Later,

4–5. To protect his wife's precious wedding ring, King George III of England gave Queen Charlotte a *keeper,* or diamond hoop. (*Photograph courtesy of Diamond Information Center*)

4–6. The five-diamond Princess-style ring was forerunner of the tall, pronged-head ring.

diamonds were set in very tall pronged heads, shanks became more open and lacy. This style gave way to filigree and the use of side stones. While many varieties of solitaire diamond settings date from World War I, the demand for this style was sluggish in the 1920s when matched sets of engagement and wedding rings became popular. After World War II, designers and manufacturers offered the public a design called *fishtail,* which was widely acclaimed because a small center diamond bead-set in an illusion head made the diamond look more important, and the sides set with diamond melee gave an expensive look to the set.

It may be of interest to the appraiser that the average price of diamond engagement rings in the United States has more than doubled during the last decade. The Jewelry Industry Council reports the reason is due to revival of the romantic wedding traditions after the 1960s counterculture movement away from custom. Moreover, they say that 75 percent of all brides in the United States receive a diamond engagement ring. The most popular styles presently are the channel-set and pavé-set mountings. A return to the use of a matching wedding ring for grooms has been noted by industry experts. The message to appraisers is clear: when a client brings a set of wedding rings to be appraised for replacement, be certain to point out the need to include the man's ring.

Although the history of rings is important and interesting to appraisers, it is infinitely more practical to be able to view historic pieces, handle them if permitted, and be able to recall some firsthand information gained by the examination for use in practical appraisal techniques. Auction houses and art galleries will permit handling, and art museum exhibitions can afford the opportunity to view but not touch. However, the appraiser needs to be aware of any opportunity to survey and study jewelry items from the past. The money and time invested will be slight compared to the knowledge gained.

The appraiser examining rings for valuation is subjected to a wide range of styles, both modern and antique. The rings may be unadorned metal or set with a variety of natural gemstones, synthetics, imitations, glass, enamels, or organic materials. Hundreds of classifications exist: ecclesiastical, signet, presentation, Masonic, fraternal, fraternity, military, heraldic, love, betrothal, marriage, posy, gimmal, mourning, memorial, hair, portrait, inscription, historical, commemorative, religious, charm, magic, medicinal, poison, ornamental, puzzle, key, watch, dial, compass, and more.

Regardless of the age, style, or motif, the valuer should remain confident in the use of standard appraisal concepts and principles and the methodology for research and valuation expressed in this book.

Common Elements in a Ring That Can Be Compared

If using the market data comparison approach to estimate value, certain elements in a piece of jewelry can be used for comparison to another of its kind. A mar-

4–7. Common elements in a ring that can be compared: metal, manufacture, style, condition, period, gemstones, and provenance.

ket data comparison analysis works for subjects for which there are comparable properties that have sold in a market. To value one-of-a-kind or irreplaceable items, use logic and reasoning and associate values. Look at historical price for comparables; note differences and similarities as a measurement for determining value. Use of cost approach *and* market data approach may be needed.

The following elements can be used to compare the rings in figure 4-7. A scale can be set up by the appraiser to see how the items rank. This kind of comparison scale and ranking system can be used with all jewelry where the market data comparison approach is used.

Compare and Rank:

1. Type of mounting manufacture
2. Style or motif of mounting
3. Care of finish of mounting
4. Metal fineness
5. Weight of metal
6. Condition of mounting
7. Period of manufacture
8. Wearability
9. Demand
10. Size and weight of gemstones (if any)
11. Quality of gemstones
12. Condition of gemstones (chipped, pristine?)
13. Other decoration, that is, enamel, engraving
14. Provenance

Determining Appropriate Market

Jewelry evaluation has recognized levels of markets appropriate for specific purpose and functions. They are termed *most appropriate market* on the report and vary depending upon age, condition, quality, intrinsic elements, aesthetic appeal, and provenance of the jewelry item. Also, other factors may influence the *most appropriate market* such as fashion, period of manufacturing, creative interpretation, and so forth.

Three markets used as *most relevant and appropriate* when estimating value for fair market value are: scrap, auction, and retail jewelry.

- *Scrap.* Jewelry unmarketable in its present state and in which the value is primarily the intrinsic content would be consigned to scrap.
- *Auction.* A market for like-item jewelry in a used but nearly new condition would be the auction. This market is selected for jewelry items that have appeal to collectors, exhibit a universal interest as antique or period items, were made by a recognized designer, have an important jewelry retailer name attached, come from a famous collection, have historical attachment or provenance, are considered rare in the market, or have an outstanding element of quality.
- *Retail Jewelry.* If the appraiser selects this market as *most appropriate,* the item should be in new or nearly ideally the item should be in new or nearly new condition, and if not, it should be saleable and in demand.

The following examples of appropriate Fair Market Value (FMV) for various jewelry articles may give better insight into the various options:

Platinum, diamond, and ruby ring (fig. 4-8) in good condition. One emerald-cut diamond of 2.40 carats is set on a stem of four baguette diamonds, bordered by 12 marquise-shaped rubies and 8 round diamonds, accented by 2 diamond baguettes set on the band. *The most common retail market for FMV: Auction.*

4–8. The most appropriate resale market for this ring would be auction.

4–9. A worn wedding ring would probably be sold for scrap.

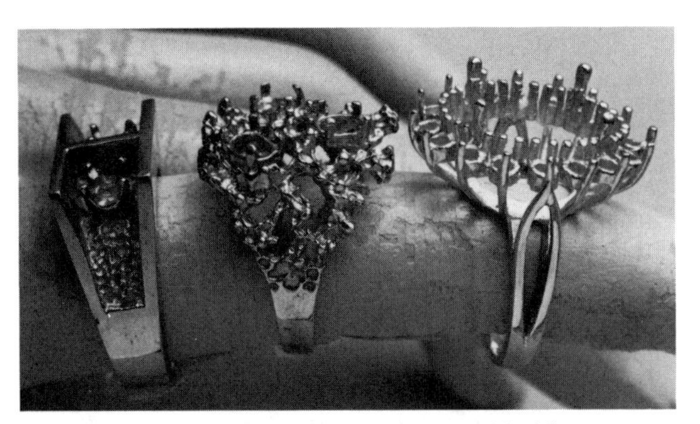

4–10. Three used jewelry mountings could either go to a dealer of secondary jewelry or be sold as scrap.

4–11. A diamond solitaire in new condition could be resold to a retail jeweler or could go to auction. (*Photograph courtesy of Diamond Information Center*)

A worn gold wedding ring (fig. 4-9), 14K gold, 4 dwts., with an inscription in the shank. *The most common retail market for FMV: Scrap.*

A group of three ring mountings (fig. 4-10), all 14K gold, yellow and white. Used mountings in an estate appraisal. In worn but good condition with a total gold weight of 10 dwts. *The most common retail market for FMV: Dealer* in used or secondary jewelry; or sell as scrap for melt value. The choice will depend upon the circumstances, condition of individual mountings, and existence of market demand for the items.

A diamond solitaire gold ring in the current style and in as-new condition (fig. 4-11). *The most common retail market for FMV: Retail Jeweler, or Auction.* Other markets that may be investigated for possible use are wholesale, collector specialty market, dealer-to-dealer, museum, and antique dealers.

Writing a Description of a Ring

After the identification of the jewelry and gemstones, quality grading, and research for a value conclusion is completed, the item must be completely described on the appraisal report. The following narrative description is a Model Appraisal Narrative for the ring pictured in figure 4-12:

4–12. Gold ring with diamonds, rubies, sapphires, and emeralds.

MODEL APPRAISAL NARRATIVE

Writing a Description of a Ring

One (1) ladies stamped and tested 18 karat yellow gold ring containing twelve (12) diamonds, twelve (12) rubies, ten (10) sapphires, and four (4) emeralds.

The 18K mounting is yellow gold and of cast manufacture. Heads for gemstones have been cast in place. The shank and top is of twisted wire design with a six (6) wire split shank. The ring has an arabesque motif.

The twelve (12) full-cut round brilliant diamonds in the ring are set eight (8) in a center rosette design, four (4) at the corners of the ring. Rubies, sapphires, and emeralds are set in flower (rosette) designs on four sides of the center diamond rosette.

The diamonds are:
One (1) full-cut round brilliant diamond measured by Leveridge gauge 6.50mm x 3.90mm and weighing approximately *1 carat* estimated by standard Gemological Institute of America formula.
Clarity grade: VVS-1
Color grade: G
The diamond is set in a 7-knife-edge-prong setting.
Detailed plot analysis documented on page 7 of this report.

Eleven (11) full-cut round brilliant diamonds, 3.0mm x 1.9mm each by Leveridge gauge measurement. The weight of each diamond is approximately *0.10 carats* estimated by standard GIA formula, with the combined total diamond weight *1.10 carats.*
Average clarity grade: VVS-1
Average color grade: G
Each diamond is set in a 3-knife-edge-prong setting.

Twelve (12) natural rubies. Each is round, faceted, and 3.0mm diameter. Depth was estimated using microscope and table gauge. Each ruby is estimated to be approximately *0.15 carats* with a combined and total ruby weight of approximately *1.80 carats.*
One ruby is chipped.
The rubies are slightly purplish Red stones with medium to medium dark tone; light to moderately included with fair cutting, and approximately 40–60% brilliancy.
Each ruby is set in a 3-knife-edge-prong setting.

Ten (10) natural sapphires. Each is round, faceted, 3.0mm diameter. Due to setting restrictions, the depth was estimated by using a microscope and table gauge.
Each sapphire is estimated to be approximately *0.15 carats* with a combined and total sapphire weight of approximately *1.50 carats.*
The sapphires are violetish Blue with medium to medium dark tone; free from inclusions under 10x magnification. The stones have fair cutting and approximately 30–60% brilliancy.
Each sapphire is set in a 3-knife-edge-prong setting.

Four (4) natural emeralds, round, cabochon-cut stones are the center stones in each "rosette."

The emeralds each measure 4.6mm diameter by Leveridge gauge, and heights were estimated by using microscope and table gauge.
The emeralds are approximately *0.35 carats* each with a total and combined approximate emerald weight *1.40 carats.*
The emeralds are slightly bluish Green with medium to medium dark tone, light to moderate inclusions, and fair cutting.
The stones have a 30–50% brilliancy.
Each emerald is set in a 6-knife-edge-prong setting.

The total and combined gem weight: *6.80 cts.*

Net gold weight of mounting, 10 dwts.

The ring is size 7 with a shank that tapers from 4mm center back to 9mm at the shoulders.

The ring is approximately: 1⅛" h. x 15/16" w. x 1¼" d. (28mm x 23.5mm x 32.1mm).

The ring is trademarked 18K HB.
HB is the trademark of Hammerman Bros, Inc.

GIA or AGL Color Grading System Applied.

Current Replacement Value $________________

____
Appraiser

Questions and Answers About Rings

Q. What do the marks inside some old rings mean and how can they be read (see fig. 4–13)?

A. These are hallmarks and, for the rings pictured, British hallmarks. The marks define the fineness of the metal assayed plus an entire series of other informational marks. A British hallmark usually is comprised of five marks: the manufacturer of the article, the mark defining the precious metal content of the alloy, the karat mark of the metal, the Assay Office mark to identify the office in which the article was tested and marked, and the date letter that indicates the year in which the article was hallmarked (not necessarily in that order). There are numerous books that offer detailed information about hallmarks. Two are *Guide to Russian Silver Hallmarks* by Paul L. Paulson and *Bradbury's Book of Hallmarks.*

Q. What if the hallmark or maker's mark looks like it might be special but cannot be found in any available book on the subject?

A. If the books at your disposal have proved to be of no help and you have exhausted your resources, you may want to footnote that fact in your appraisal report. You may also wish to reproduce the marks (to the best of your ability) as archival documentation. This could be of value to a future appraiser who may have some hallmark information through special channels or books. Many unidentified marks of the past are being researched by a new generation of appraisers using advanced technology and research methods. Some marks are turning out to be pseudomarks used to fool the unwary.

Q. What do the words "solid gold" found on the inside of some old rings actually mean (see fig. 4–14)?

A. This is a term some would reserve for 24K gold that has no other elements in the metal. However, on June 13, 1967, the Federal Trade Commission in an advisory opinion held that "solid gold" may be used to describe items that do not have a hollow center and have a fineness of 10 karat or more.

4–13. Examples of British hallmarks.

Earrings

The history of earrings covers hundreds of cultures and crosses all social barriers. The custom of wearing earrings as ornament is probably of ancient Oriental origin as far as it can be traced. Earrings of outstanding goldwork have been discovered during archaeological excavations of primitive societies. Ice Age rock drawings along the Spanish Mediterranean coastline depict nude human forms covered in colorful ornamentation, including earrings.

From the Bronze Age to the Atomic Age, earrings have been part of the jewelry costume of all cultures. From early recorded times, earrings have been worn by the Egyptians, Arabs, Spaniards, Romans, Greeks, Germans, and Gauls, by men, women, and children.

In Greek and Roman classical antiquity, earrings with the crescent moon were among the most popular and earliest designs. This motif was probably meant to be of symbolic significance.

Although earrings were important symbols of royalty and wealth, it is believed the original purpose was amuletic, used to protect the wearer from harm, the evil eye, disease, or to invoke favors from the gods. A deep and universal belief in omens extends to a profound connection between jewelry and religion.

In general, the use of earrings among ancient Greek men was restricted to those who lived in Cyprus. To most Greeks, a man wearing earrings was marked as a foreigner. Hoop earrings with filigree disks were popular for Greek women. In *The Odyssey,* Penelope's suitors offer jewelry, with special mention of earrings with clustering drops, to overcome her reluctance.

4–14. The ring is marked "solid gold."

Roman women wore three or four earrings on each ear, some of thin coils of gold wire or gold foil decorated with precious stones. The Romans developed designs using large colored stones at the center. They multiplied the species of stones, set them in rows, and bordered them with pearls. The Romans used the processes of hammering, chasing, casting, repoussé, engraving, welding, and soldering. The art of granulation reached its apex in the seventh and eighth centuries followed by a decline of this technique.

In the Renaissance during the fifteenth and sixteenth centuries, earrings along with other jewelry received tremendous stimulus. The earrings of the period are characterized by the rich color used in enamels and colorful gemstones. Earrings became almost exclusively the domain of women during the Renaissance. Pendant styles set with precious stones were a perfect jewelry accompaniment to the elaborate coiffures and jewel-encrusted gowns. In Italy religious motifs were forsaken for the classic and naturalistic look that swept France, Germany, and the rest of Europe. The elaborate earring designs seen in paintings of Holbein, Dürer, and Rubens set the trend.

Men have persisted in wearing earrings, one or two, since Elizabethan times (1558–1603). Sir Walter Raleigh wore pearl earrings, as did Shakespeare and many of his contemporaries. Henry III of France wore a long, dangling pearl or precious gemstone earring in each ear. Charles I, always fashionable, wore his ⅝ inch-long dangling pearl earring to the scaffold in 1649. The earring is said to be preserved in a private collection hidden away in England to this day.

During the reign of Louis XIV (1643–1715), French jewelry was considered the best designed and French craftsmen the finest in the world. Society was not disappointed in a special design called *girandolé* (fig. 4-15). This was a pendant earring with three small precious stones or pearls suspended from a main disk, crescent, ribbon swirl, or horizontal bar. The girandolé design was immediately embraced by the fashion conscious women of the period. The precursor to this style was the Roman earring known as a chandelier style. The popularity of girandolé style made it the harbinger of an even longer pendant earring called a *poissardes* earring (fig. 4-16) in the nineteenth century. These excessively long earrings, often falling from the ear lobe to rest just above the shoulder, were perfect accents for the plunging décolletage of the fashionable lady. To gild the lily further, the poissardes earring was enhanced with briolette-cut gemstone. The briolette-cut was popular because of its all-around faceting of the gemstone, which produced scintillation and light

4–15. The *girandolé* earring style. *(Illustration by Elizabeth Hutchinson)*

4–16. The *poissardes* earring was so long it often touched the shoulder. *(Illustration by Elizabeth Hutchinson)*

dispersion with the movement of the earring. In her book, *A History of Jewellery 1100–1870,* Joan Evans writes of mid-nineteenth-century pendant earrings reaching lengths of four inches!

Men's wearing earrings prevailed as custom into the nineteenth century and the time of Napoleon. As France set not only costume but jewelry fashion, during the nineteenth century men ceased wearing masses of jewelry and settled for a ring or two and several seals on a watch fob. Fashion aside, the nonconformist is always present in society. In their book, *5,000 Years of Gems and Jewelry,* Frances Rogers and Alice Beard note: "Occasionally a fop would go so far as to wear earrings; if ridiculed for vanity, he had the excuse that piercing the ears and wearing earrings was therapeutic." Further, the authors reveal, "earrings were considered a prevention for eye diseases."

The earring has not evolved so much as it has been recycled. Earrings are closely connected to fashion and were worn in all periods when hairstyles allowed the ears to show and when the costume did not detract from the earrings effect.

Victorian-period earrings went from one extreme to the other. The early part of the period called for large, fancy bonnets for ladies, and their long curls, braids, or ringlets fell in front of, or completely covered, the ears. In the middle of the nineteenth century, the visibility of ears was once more acceptable and earrings became part of fashionable dress. These earrings were button shaped, elongated beads, or small gold hoops. Earrings went from the double or single ball drops or small flattened hoops to (once again) earrings so long that, according to writer Margaret Flower in *Victorian Jewellery,* "they began to tangle in bonnet strings." The lengthening and shortening continued along with the development of earring findings (fig. 4-17).

There are seven basic earring forms: disk (or button), spiral, boat or crescent, pendant, disk and pendant, hoop, and a baule or box form.

Common Earring Elements That Can Be Compared

Study the pair of earrings in figure 4-18 for a survey of elements that may be compared and ranked with like kind. Observe the following:

1. Metal type of mounting (gold, platinum?)
2. Type of back (clutch, screw, hook?)
3. Metal fineness
4. Metal weight
5. Type of manufacture
6. Name of manufacturer or designer
7. Condition of the item
8. Motif and period of manufacture
9. Gemstones, if any
10. Quality of gemstones
11. Size and weight of gemstones
12. Trademark or maker's mark.

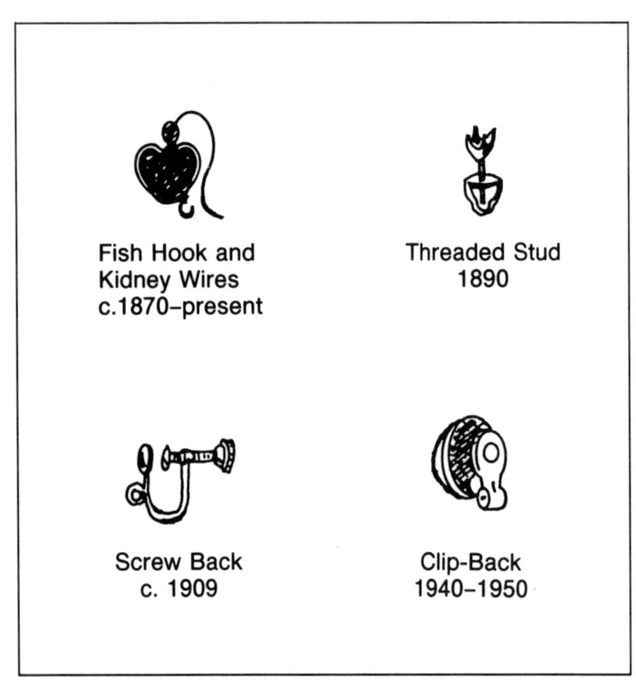

4–17. Evolution of the earring finding.

4–18. These platinum, diamond, and aquamarine earrings can be compared with like kind to estimate value. A comparable should have similar qualities of manufacture, metal, gemstones, weights, and period of design.

MODEL APPRAISAL NARRATIVE

Writing a Description of Earrings

4–19. White gold earrings with pearls and diamonds.

One (1) pair of 14 karat stamped and tested white gold earrings containing seven (7) round cultured pearls in *each* earring, and twelve (12) full-cut round brilliant diamonds in *each* earring.

The 14K white gold mounting has 14K white gold clips assembled on the backs. The mountings are of die-struck manufacture and the earrings are designed in a rosette, or flower, motif with six (6) white gold leaves as accents. Each leaf is bead-set with a diamond. Six (6) diamonds are set in 4-ball-prong stem-set heads and encircle the center cultured pearl.

The *pair* of earrings contain the following diamonds:

Twelve (12) full-cut round brilliant diamonds measuring 3.0mm x 1.9mm by Leveridge gauge and with an estimated weight by standard Gemological Institute of America formula of *0.10 carats* each.

Average clarity grade: VVS-1
Average color grade: H
Well cut

Twelve (12) full-cut round brilliant diamonds measuring 2.4mm diameter by Leveridge gauge. Depth cannot be measured due to mounting restrictions. Estimated weight of the diamonds is approximately *0.05 carats* each.

Average clarity grade: VS-1
Average color grade: H
Good cut.

Total combined diamond weight in the *pair* of earrings approximately: *1.80 carats.*

The pair of earrings contains a total of fourteen (14) round cultured pearls; two are 8.5mm each and set in the center of the flower design. The cultured pearls are spotless with high luster and medium nacre. Twelve (12) round cultured pearls, 7mm each, are set around the center pearls. They are slightly spotted with high luster and medium nacre. All pearls have a white body color with rosé overtone and are well matched.

Total weight of the earrings: 10.50 dwts., including cultured pearls and diamonds.

There is no trademark or maker's mark.

Current Replacement Value $________________

____
Appraiser

4–20. If sent to auction, the reserve price for these earrings would be set by the auction representative and agreed to by the seller.

Questions and Answers About Earring Appraisals

Q. I am sending some estate earrings (fig. 4-20) to auction. What is the reserve price, and how is it determined?

A. A reserve price is the confidential minimum price agreed upon between the seller of the item and the auction house, below which the jewelry will not be sold. Auction spokespersons acknowledge that the reserve is about 60–70 percent of the auction catalog pre-sale estimated figure.

Q. I have a pair of multi-diamond earrings to appraise. If the diamonds in one of the earrings are exactly like the diamonds in the other half of the pair, is it necessary to record measurements, estimated weights and quality grades on *both* earrings?

A. For professional appraising, the answer is yes. The answer is also yes in the case of a piece of jewelry set with numerous diamond melee. Each diamond should be measured by Leveridge gauge, weights estimated, quality grades noted, and information recorded on the appraisal document.

Q. I have a pair of estate jewelry earrings for appraisal. They have been badly repaired. How much difference will this make in the estimation of value?

A. According to the experts at major auction galleries, if an item has dents or a piece broken off and has to be soldered, then it never looks the same. In that case, for valuation, it may be anywhere from 25 to 50 percent off the estimate of an undamaged piece.

Q. Do earrings have any utility or are they just an item of fashion?

A. Earrings are the only jewelry that is totally nonutilitarian and completely decorative. In some cultures they may still be considered therapeutical and/or magical. Many earrings have historical significance because of their jewels, their craftsman, or an important owner. Historical importance, *if it can be proven,* will add to the value. However, heresay from the client or unsupported opinion cannot be considered provenance.

Q. When did the practice of piercing the ears have a revival?

A. Between 1880 and 1900 earrings had a general decline in popularity. The invention of a patented screw device on January 2, 1894, that allowed ear piercing with a stud, which attached to the ears and did not mutilate the flesh, brought about a revival of the practice by 1900.

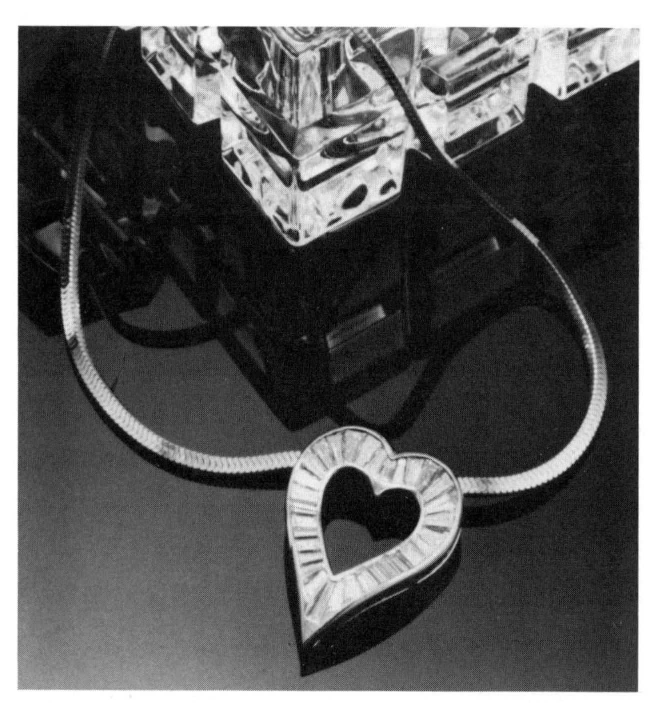

4–21. A necklace is one of the oldest body ornaments. *(Photograph courtesy of Diamond Information Center)*

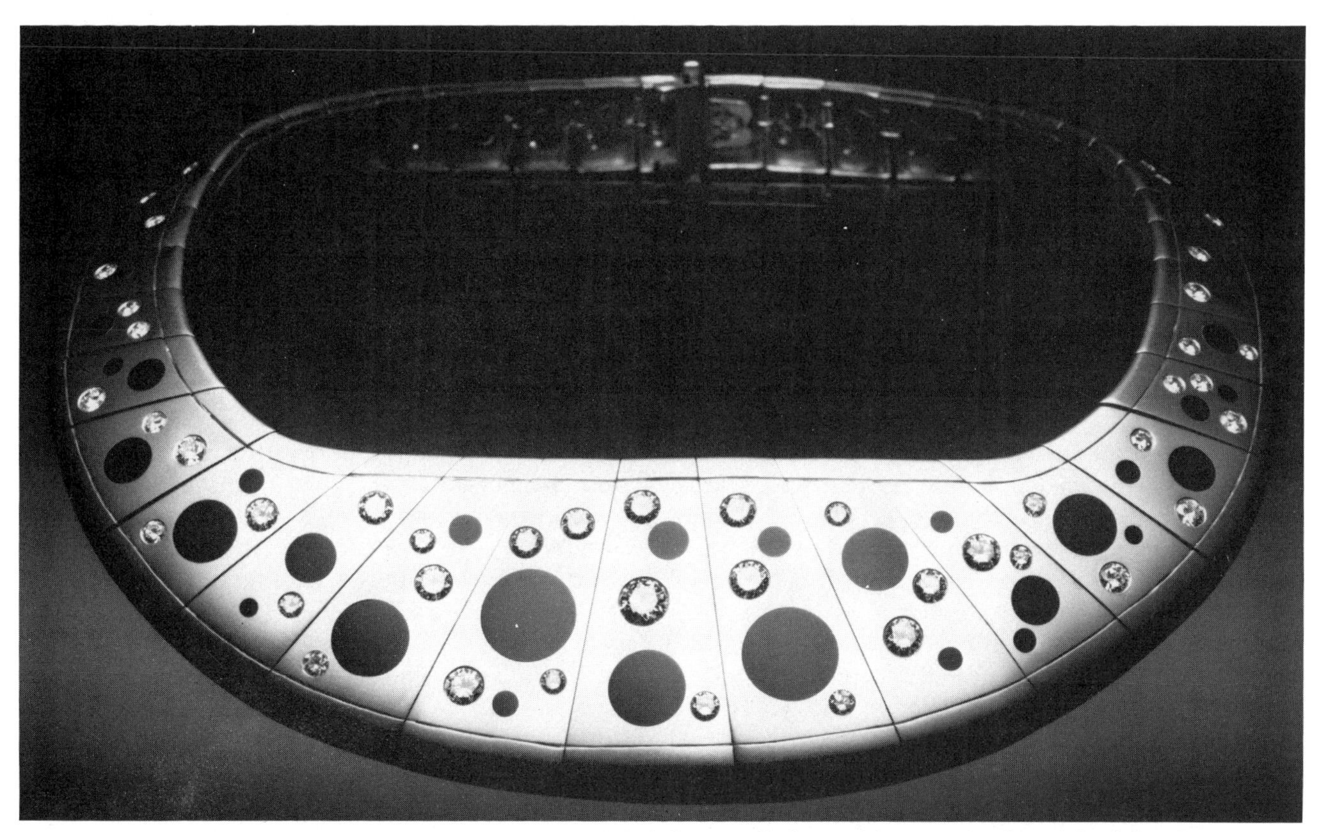

4-22. As long as it encircles the neck, the piece can rightfully be called a necklace! Design by Michael Bondanza. *(Photograph courtesy of Diamond Information Center)*

Necklaces

The necklace, necklet, or torque is one of the oldest articles used for body adornment. It can be worn alone as a metal coil (or band of cloth or fur) around the neck or with a pendant as illustrated in figure 4-21. Necklaces can be beads, a linked chain, or a graduated row of pendants (fig. 4-22) hanging in assembled linkage, whatever! As long as it encircles the neck, the item is a necklace.

The term *necklace* was first used in 1590 in the play *The Tragedy of Dr. Faustus* by Christopher Marlowe. Before this word was coined, the term *collar,* or *coler,* was the most favored expression. In countless jewelry inventories of historically famous and infamous families, there is mention of "colers." Two such necklaces played roles in the fate of England, one pawned by Henry V before the Battle of Agincourt in 1415 and the other by Henry VI to raise money for war. The one worn by Henry V was called the *Ikylyngton Coler* and was richly inlaid with rubies, sapphires, and pearls, according to the authors of *The Book of Necklaces.* The other necklace was the *Riche Coler* and is recorded as "a pesane, otherwise cleped a coler, of gold."

Throughout lands north of the Mediterranean all the way to the British Isles, ancient nude figurines and votive objects have been found adorned with at least one necklace, sometimes painted on the figure. The designs range from chokers to beads and chains, often several strands and of various lengths. In the British Museum, along with ancient relics of French cave dwellings, is a periwinkle shell necklace from a Cro-Magnon cave. This display of jewelry indicates not only the deep longing for self-adornment that often signifies power, but also the ritual importance early man attached to the necklace.

Investigation into the long history of adornment reveals that it was only in the Dark Ages that necklaces were not worn. The reasons were socioeconomic. The Dark Ages was an historical period with few vocational choices—either train for the clergy or be skilled in the sword—and neither profession had use for jewelry. The talented metalsmiths, sculptors, and artists spent their entire lifetimes in the service of decorating and ornamenting churches and needed no patronage other than the clergy. The few wealthy individuals spent no money on self-ornamentation, preferring that their riches go, via the Church, for the salvation of their immortal souls.

Not until the mid-fourteenth century did politics, society, and fashion combine to dictate low-cut décolletage, with which once again necklaces could be worn and used as bold expressions of status and power. Men and women indulged in chains and necklets lavishly decorated with gemstones. Regrettably, few medieval necklaces have survived for our personal inspection, especially those in precious metals. Most of the pieces were broken up and melted for their intrinsic value or melted down and reworked because fashion demanded a style change. Thus, we must look to paintings and tapestries for a chronicle of the styles and motifs.

Ecclesiastical jewelry was popular, with a trend toward religious motifs depicted on all types of jewelry, especially necklaces and religious medallions. These medallions were thought to preserve and protect the wearer.

During the Renaissance, opulence in jewelry exploded as men and women dressed themselves with necklaces laden with gems, as well as pendants, chains, and strings of beads and pearls. Not only was jewelry a separate accessory but it was frequently sewn onto the garment. In portraits of Henry VIII of England, one can see lavish collars or necklaces of gold, decorated with pearls and rubies, as well as gold chains with pendants. Gemstones decorate his hat, sleeves, and doublet as well.

In the seventeenth century, handsomely bejeweled necklaces set with large paste stones or foiled stones were admired. The paste jewels were faceted both front and back, even though the backs of the settings were closed. Ribbons were used on each side of the necklace and tied at the back of the neck to whatever length the wearer might prefer. The settings holding the jewels are very distinctive, with clawlike serrated points to press against the fragile glass. It was not until the nineteenth century that modern settings with open backs were introduced into the jewelry market.

A public demand for inexpensive gold jewelry led to the invention of a substitute for gold called pinchbeck, after its inventor Christopher Pinchbeck. He was a British clock and watchmaker who wanted to offer his customers inexpensive watches that looked like gold. Pinchbeck is an alloy of 83 percent copper and 17 percent zinc. The novelty caught on quickly. The substitute metal that looked like gold was so popular that its use has survived into this century. The result is that any English jewelry, especially neckchains, lockets, and watches should be carefully examined, assayed, and validated as to metal content and fineness, even if hallmarks are present.

Necklaces took on a sentimental and distinctive quality during the Victorian era (1837–1901). They were no longer acquired as just beautiful ornaments. The necklace became a vehicle for parading one's wealth and social status. Further, there evolved strict rules of etiquette and social codes of wear. An 1821 fashion magazine declares that pearl necklaces are for the virginal debutante, and diamond rivieres are for dowagers. Coral, amber, and garnet necklaces are considered appropriate for children, and the grieving widow is assigned jet. Emeralds and amethysts are only suitable for married women, and paste is good enough for actresses! All, however, were urged to advertise their spiritual allegiance by wearing a gold cross on a chain.

Necklaces from the Art Nouveau and Art Deco periods will offer the appraiser few identification or evaluation challenges because the designs are unique and auction catalogs can provide an abundance of comparables for value estimation. A late twentieth-century style that may defy appraisers, however, is that entitled Art Jewelry. These are items (figs. 4-23 and 4-24) made by jewelers who think of themselves first as artists, second as jewelers. The articles may be in gold or silver, but just as often titanium and niobium. The issue here, according to the artisans themselves, is the look, style, and message. Many artists producing Art Jewelry use diamonds or colored gems in their designs,

4–23. Wearable art jewelry by Kevin Lane Smith of Seattle in a 14K, 2½-inch-long pendant with Oregon Opal. His extraordinary work concentrates on revealing color and light transmitted through gemstones. The pendant pictured is $1,600. (*Photograph courtesy of Kevin Lane Smith*)

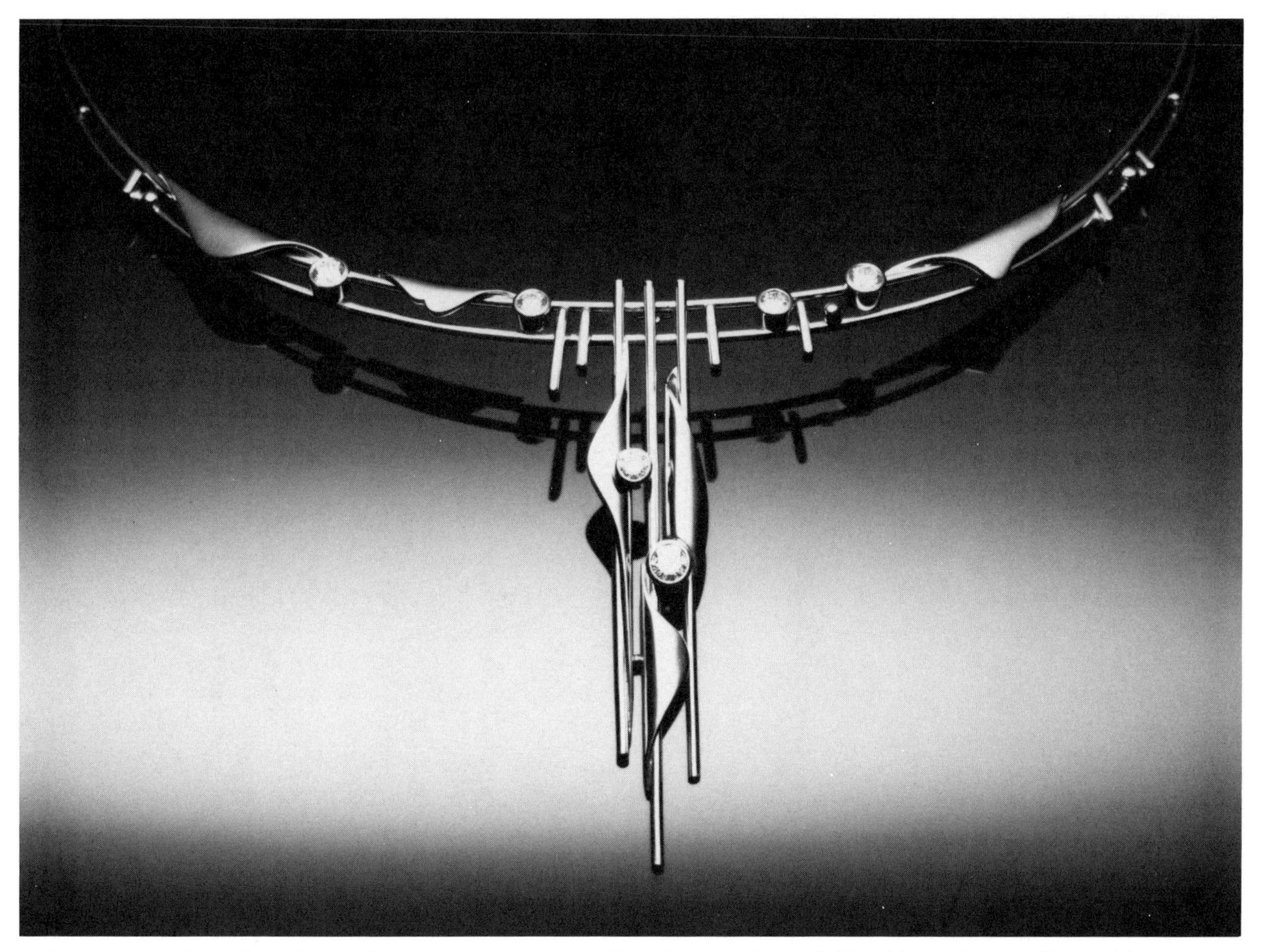

4–24. A carat of bezel-set diamonds appears as staccato points along the bars of this whimsical 18K yellow gold piece with sandblasted finish. The piece was designed by German artisan Lothan Kuhn. (*Photograph courtesy of Diamond Information Center*)

but some treat the metal to create colored surfaces. The technical skill brought to this art is astounding and the demand for this kind of crafted jewelry is growing. Consumer and trade magazines such as *Apollo, Lapidary Journal,* and *Ornament* spotlight trends and stimulate public awareness and desire for the work. The Springfield Craft Fair located in Massachusetts is a spawning ground for many of these talented artisans. The Designer Room at the Jewelers of America New York show always displays articles by these craftsmen. Those articles appearing at the New York show, in fact, have had significant influence on mass-market trends.

Appraisers should take heed of this growing segment of the industry and become better acquainted with those special factors that attend the valuation of such jewelry, such as detail in design, exotic metals, limited numbers of the work, and extraordinary metalsmithing.

Questions and Answers About Necklace Appraisals

Q. What is "French jet?" Is it (fig. 4-25, see p. 68) actually the material jet?

A. The name "French jet" originated in France about 1820 when the French began to make black glass that could be used in mourning jewelry. Jet came into fashion in 1820 when the assassination of the Duc de Berry plunged society into mourning. Real jet, a natural mineral, was mined at Whitby, England, and was used through the decades of the century when Queen Victoria's widowhood kept mourning in fashion in England. French jet is a glass substitute for the real mineral jet. Jet is no longer mined in any quantity in Whitby.

4–25. A French jet necklace.

4–26. Daguerreotype lockets can be dated circa 1840.

Q. I wish to circa date a daguerreotype locket necklace (fig. 4-26). How old is this type of jewelry?
A. The first lockets with daguerreotypes were made in the United States about 1840. After the camera was invented, photos replaced painted portrait jewelry.

Q. I have a silver necklace that is marked "Taxco" on the back. What does that mean?
A. It means that the piece was made in Taxco, Mexico, and is probably handmade.

Q. Does a cast solid 14K gold necklace with imitation gemstones fall into the costume or fine jewelry category?
A. It can be classified as a piece of commercial quality jewelry.

Q. I am appraising a carved and decorated ivory tusk pendant. Sailing ships and whales are part of the design. How should I advise my client in its care?
A. Your client has a piece of scrimshaw folk art. The carving of scrimshaw dates back to nineteenth-century whaling ships. The ivory may be walrus tusk. Do not wash it with soap and water or use any kind of oil on it as that would change the color of the ivory and lower the value.

MODEL APPRAISAL NARRATIVE

4–27. Diamond and ruby necklace with detachable pendant.

One (1) 14K stamped and tested white gold chain with white gold detachable pendant mounting containing six (6) rubies, forty (40) full-cut round brilliant diamonds, twenty-two (22) marquise-cut diamonds, and ten (10) baguette diamonds.

The necklace is 24 inches in length, including clasp, and the chain portion is 2mm wide and designed in the infinity motif. The necklace terminates in a 14K white gold barrel clasp with double figure *8* safety catches. The necklace has a 14K white gold center section that is a detachable pendant in a flower-and-leaf swag motif. The mounting is of die-struck manufacture with assembled die-struck heads.

One (1) pear-shaped ruby is suspended from the center of the pendant; five (5) round faceted rubies are set into the center of five diamond flower designs.

All rubies are natural, transparent, and matched in color. The rubies are:

One (1) pear-shaped ruby, slightly purplish Red, medium to medium dark tone, light to moderate inclusions with fair cutting, and approximately 30 to 50 percent brilliancy. GIA or AGL color grading standards used. Plot on page _____ of this report.

The ruby measures 6.1mm x 8.2mm x 3.9mm by Leveridge gauge and has an estimated weight by standard GIA formula of *1.40 carats.* The ruby is set in a 4-barrel-prong wire head.

Five (5) round, faceted natural rubies, slightly purplish Red, medium dark tone, moderate inclusions with fair cuttings, and approximately 30 to 50 percent brilliancy. GIA or AGL color grading standard applied. Each ruby measures 4.5mm x 3.5mm by Leveridge gauge and has an estimated weight by standard GIA formula of approximately *0.50 carats.* The total and combined estimated weight of the five rubies is approximately *2.50 carats.*

Each ruby is set in a 4-barrel-prong head.

The necklace contains forty (40) full-cut round brilliant diamonds each measuring 2.9mm x 1.85mm by Leveridge gauge, with estimated weight by standard formula of *0.09 carats,* and total estimated weight *3.60 carats.*

Average clarity grade, VVS-1
Average color grade: G
Good cut and finish
Each diamond is set in a 4-barrel-prong head.

Twenty-two (22) marquise-cut diamonds measuring 3.5mm x 1.7mm x 1.5mm by Leveridge gauge, with estimated weight by formula *0.05 carats* each and with total estimated weight *1.10 carats.*

Average clarity grade, VVS-1
Average color grade: G
Good cut and finish
Each diamond is set in a 4-barrel-prong head.

Ten (10) baguette diamonds measuring 3.5mm x 2.0mm x 1.4mm by Leveridge gauge and with estimated weight by formula of *0.08 carats* each, and total estimated weight *0.80 carats.*

Average clarity grade, VVS-1
Average color grade: G
Good cut and finish
Each diamond is set in a 4-barrel-prong head.

Combined total diamond weight in the necklace is approximately *5.50 carats.*

Necklace total weight, 17 dwts., including diamonds and rubies.

The necklace has no trademark or manufacturer's mark. The barrel clasp on the chain is marked "Italy."

Current Replacement Value $_______________

____
Appraiser

Pendants

Pendants are among the most beautiful of ornaments and their styles and motifs are manifold. Pendants can and do vary in size, style, and design and are normally worn suspended from a chain or cord. Both ancient and modern pendants can be used as objects to amplify the beauty of a neckchain or as solitary statements of fashion. Examples of pendants include lockets and crosses. Shapes range from oval, round, oblong, heart, and rectangular to any outline imaginable. Pendants may be constructed of precious metals, gold-filled, gold-plated or base metal, and may be set with gemstones or enamelled. They may be frames containing miniature portraits or lockets housing objects of veneration or coins.

Laws Forbidding Wear

In her book *An Introduction to Courtly Jewellery,* Anna Somers Cocks points out that the history of much English jewelry can be traced to passage of the sumptuary law in 1363 that regulated how much the private citizen was allowed to spend on luxury items. It was passed during the reign of Edward III because the king and his court felt that too many of the "wrong sorts" (commoners) were buying and wearing jewelry. With so much glitter around, the lawmakers felt the innate differences of rank and social standing were not readily apparent. The law forbade the wearing of belts, collars, clasps, rings, garters, brooches, ribbons, chains, bands, seals, and anything of gold or silver by handcraftsmen, yeomen, and their wives and children! A form of jewelry, the pendant, was not listed in the sumptuary law because it did not exist in that cultural period. By the 1500s, however, the pendant appears in the sketchbooks of pendant pattern designs from Hans Holbein produced for Henry VIII of England. In the designs of Holbein, we can see the depth of skill and breadth of imagination that defines the typical Renaissance craftsman.

Pendant History

Usually the pendant has no direct function to fulfill except that of bodily adornment. It does not hold together the costume as can a brooch nor does it set a seal upon an important document as does a seal ring. Its genesis is in the amulet or talisman to which a magic power was attributed. The pendant has been worn on a ribbon, small chain around the neck, on the belt of a garment, on a bracelet as a charm, on a watch strap as a fob, or on a hat or cap. Pendants ornamented with colored gemstones, enamels, niello, filigree, granulation, and portraits have all been found among the treasures of the Roman Empire. Most of them have an astonishingly contemporary design.

When the pendant as popular jewelry was in favor, men, women, and children all wore it as adornment. Many valuable examples set with precious stones, pearls, and enamels survive from the Renaissance. Some of the finest examples can be viewed in the Walters Art Gallery in Baltimore. Except for a few of the most famous designers, it is impossible to attribute the jewelry produced during the Renaissance to a specific craftsman or country of origin. Jewelry craftsmen were called to work at the great royal courts and stayed until their patrons died or got tired of them; then they moved to another court. So goldsmiths were itinerant workers wandering from place to place. One work, however, is so distinctive that it can be accurately attributed to German craftsmen. The pendant is called a *Gnadenpfennig* and was molded to show the head of a ruler. It was worn on a chain. Other popular motifs were pear-shaped forms with hanging pearls and designs with Christian figures and scenes. Greek mythology and animal motifs, birds, and legendary beasts were common.

For a short period, pendants were known as *decorations* and used as part of secular and ecclesiastical costume during special occasions. This practice was discontinued in the seventeenth century in favor of wearing a solitary cross on a chain.

In the Georgian period, necklaces with diamond pendants were fashionable. They were worn around the neck on a black velvet ribbon and accented the plunging décolletage of the period.

During the last half of the nineteenth century, the upper social classes of the new industrial-commercial society were fond of displaying their riches by wearing jewelry with diamonds and other costly stones. Diamond-set pendants were popular advertisements of wealth. About this time, however, the pendant lost its independence and merged with the necklace.

Sylvie Raulet in *Art Deco Jewelry* writes that the most important piece of jewelry in the late 1920s was the pendant. Suspension from an extra long chain was the most prevalent way to wear one, but pendants were also worn pinned to lapels and collars.

Modern pendants offer wear options. Most have large enough bails that they can be removed from their chain or cord for other use. Some pendants are also brooches and have both bail and pinback. The appraiser should consider the pendant as a separate item when writing a valuation unless it is permanently attached by integrated chain assembly into the necklace.

The following photographs illustrate a pictorial evolution of the pendant: figure 4-28, antique pendant circa 1880; figure 4-29, estate item circa 1940; and figure 4-30, a smart contemporary design of the 1980s.

4–28. An antique gold pendant, circa 1880, estimated value $400. The shield-motif pendant incorporates many of the aforementioned design elements: torsade border, pateration, and cannetillé.

4–29. A 14K white gold and diamond pendant, circa 1940, estimated value $350.

4–30. (*right*) A dazzling contemporary-styled diamond pendant with sculptural look and specially made chain. (*Photograph courtesy of Diamond Information Center*)

MODEL APPRAISAL NARRATIVE

Writing a Description of a Pendant

4–31. An heirloom pendant with enamel-on-gold paintings and table-cut and rose-cut diamonds. (*From the Epstein Collection*)

4–32. The pendant gives no indication of the hidden painting under the cover. (*From the Epstein Collection*)

4–33. Another compartment, this one for two photographs. (*From the Epstein Collection*)

One (1) three-part locket/pendant with two enamelled portraits and an inner compartment with frames for two photographs. The pendant is set with twelve (12) diamonds in a combination of table-cut and rose-cut styles, in individual rosette-cup heads with rolled-over gold holding them in place. "S" scrolls are assembled on the top of the pendant for embellishment.

The oval-shaped pendant has been tested as 18 karat yellow gold with a 14 karat self-bail. It measures 1½″ high x 1″ wide x ½″ deep (36.5mm x 29mm x 9.4mm). The pendant is a combination of fabricated and hand-assembly manufacture, circa 1860.

The 12 table-cut and rose-cut diamonds have a combined and total estimated diamond weight of approximately *0.20 carats* (1/5 carat).

The portrait on the top of the pendant is enamel on gold in a rectangular frame with a torsade border. The enamel has had restoration in the upper left-hand corner of the portrait. Quality of the restoration is fair.

The quality of the painting is very good and typical of the mid-nineteenth-century artisans. The subject is a woman in classical dress, flowing garment with swirling ribbons and folds. She rests one arm on a leaf-entwined column while reaching skyward with a gracefully turned arm and hand. Her mood is somber; she may be an allegorical figure.

The framed portrait is on a hinge that opens to reveal an inside portrait of a reclining nude female. The nude is adorned with a bracelet and diadem, with slave bracelets on the ankles. Musical instruments and fruit are at her side. Water flows from a spigot beside her and the background is a lush foliage scene. The medium is enamel on gold.

The third compartment has an oval space for two photographs. The back of the pendant is solid and without decoration. It has some minor surface dents. No maker's mark or hallmark appears on pendant.

The pendant weighs 11 dwts.

Current Replacement Value $________________

____

Appraiser

Questions and Answers About Pendants

Q. I have a three-color gold pendant. Do the three colors of gold increase the value?
A. A three-color gold jewelry item is trendy, and the desirability might be increased as well as the value. There is extra manufacturing and labor charge for a three-color gold item, so the value may be slightly more than its identical counterpart in a single-color gold.

Q. I have a pendant for appraisal with a large center stone, and I think there may be a chip on the stone but hidden under a prong. How should this be written on the appraisal report?
A. If the prong can be lifted during the examination with the client's permission and without damage to the stone, the appraiser should try to verify suspicions. If the prong cannot be lifted and a chip or nick is suspected, the appraiser should state that in the Limiting and Contingent Conditions portion of the report by inclusion of the condition: "Gemstones examined only to the extent permitted by their mountings. No gemstones were removed from the mountings during the examination for an exact quality grade or weight."

Q. If a heart-shaped diamond pendant has a nick in the diamond, is the stone appraised using the recut weight?
A. It depends upon the size of the problem. No, if only a tiny nick exists, since it will not make a big difference in the overall weight. Yes, if it is a sizable chunk out of the stone. Figure the recut weight before estimating a value for insurance. However, if the piece is being appraised for fair market value in an estate, such as probate, appraise it in an "as is" condition.

Q. What is the best way to count baguettes?
A. From the reverse side of the mounting. It is usually easier to make weight estimation too.

Chains

The common reward for faithful service to a prince or noble in the fifteenth and sixteenth centuries was a present of a gold chain. This was among the most common type of jewelry worn by men and women, and the trend was to wear it displayed either around the neck or draped over the shoulders. Portraits of stylish men and women of this period are testimony to the fad.

Chains were frequently used and worn as badges or signs of political office like a mayoral chain of office, or as a symbol of brotherhood and orders of chivalry. Most chains were gold, silver, or silver over base metal and made with individual "S" links.

The sixteenth-century name used for chains in English jewelry was *carcanets.* This defined the large, fancy, embellished, and decorated individual links joined together (fig. 4-34). English chains with flower designs in gold and white or green enamel and white and gold roses linked with colorful leaves attest to the taste in motif and enamelling during this time. Joan Evans describes the wearing of huge gold chains by English and Italian men in the sixteenth century, while in Flanders and Germany such chains were strictly for women. The mode, Evans says, originated in Italy, but she calls attention to the drawings of Hans Mielich of Munich, circa 1570, in the Royal Library at Munich, which illustrate the vast jewelry inventory of Duke Albert V of Bavaria and Anne of Austria. Their jewelry treasury held gold chains in abundance. Interesting and important to the authentication of a carcanet is the elaborately enamelled Moorish designs on the backs of the articles.

Chains have changed little through the centuries. They have evolved from small link to massive link and back to tiny. Chains are constructed as a series of rings, links, metal beads, or disks connected one to another. A chain can be of a precious metal, base metal, or both, with size and materials varying according to the purpose for which the chain was designed. The links are usually crafted by bending a piece of wire and soldering the ends together in various styles: circular, oval, flat, cylindrical, rosette, double, triple, plaited, and so on. Some machine-made chains today are fed into the oven with solder inside the links and need no

4–34. Two styles of *carcanets,* a sixteenth-century name for chains. (*Illustration by Elizabeth Hutchinson*)

further soldering during their manufacture. This technique is helpful to the bench jeweler if later repairs are needed, because the chain (including hollow chain) may now be fixed in an efficient way. Stiff spots in a chain can result from inexperienced repairing when solder is overheated. Poorly made repairs reduce any chain's value.

The task of the craftsman has always been to treat the metal in such a way that will render it flexible but unyielding to whatever amount of tension the wearer may apply. If an appraiser is examining an ordinary chain that was mass-produced, the flexibility of the chain should be noted along with the length, metal content, weight, style, and clasp. The appraiser should not place undue value emphasis on the design since the chain has mainly a functional role. Unless, as in some fancy-motif and stylistically designed handmade chains, ornamental design can be a major factor in the artistic effort. Then it should be carefully considered in the valuation estimate.

Most chains close with some type of catch. The appraiser should inspect the findings closely to determine whether it opens and closes easily and remains closed after it has been positively shut.

Questions and Answers About Chains

Q. Chains of 45 inches in length or longer frequently turn up in estate and antique appraisals. What was their use and how were they worn?

A. Chains 45 inches or longer were often used as muff chains in the early nineteenth century. The chain was attached to each end of the muff and then placed around the neck. A muff was a material or fur handwarmer/purse accessory. Also, long chains were often worn alone as jewelry, and the style was to wear them diagonally across the body from shoulder to waist when they were popular in the Victorian era.

Q. What is a bookchain?

A. A book-link chain is the name of a style of chain with alternating round or oval links and flat plaque-links popular about 1880, often copied in revival jewelry and seen in reproductions today.

Q. What is the best way to polish a chain?

A. If polishing with a rotary tool, be sure to wrap the chain around a polishing board before beginning, or the chain could wrap around the arbor and injure the hand.

MODEL APPRAISAL NARRATIVE

Writing a Description of a Chain

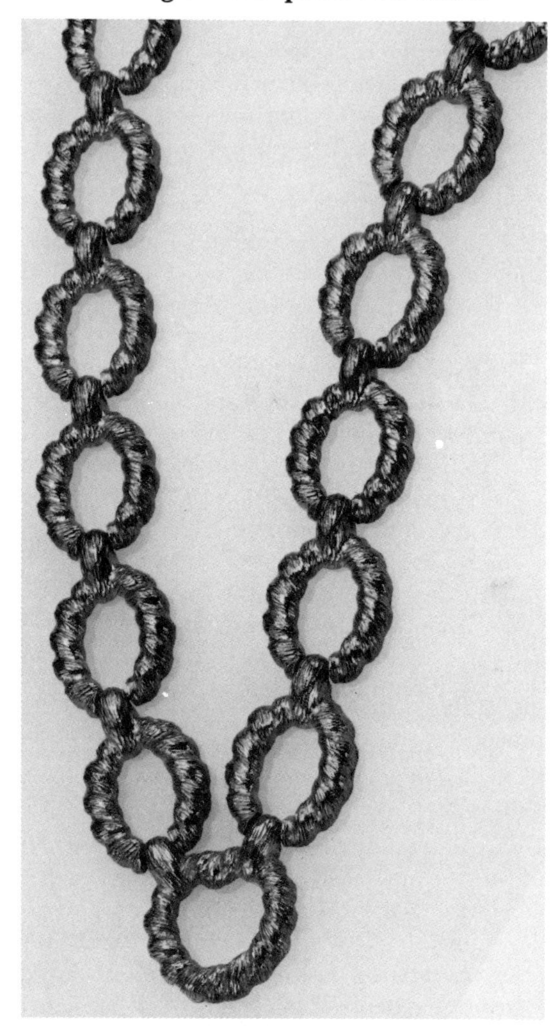

4–35. Gold neckchain.

One (1) yellow gold neckchain, stamped and tested as 18 karat. The chain is 22 inches long, including clasp, and designed with open oval-shaped fluted links and has a link that is also a hidden clasp.

The links are hollow and machine manufactured.

The fluted links have a florentine finish.

The chain's total weight is 45 dwts.

The chain is marked with the name of the manufacturer: Van Cleef & Arpels.

Current Replacement Value $______________

____

Appraiser

Bracelets

The bracelet, now almost an exclusive accessory for women, was also formerly worn by men, not on the wrist as is the present custom but high up on the upper arm. This custom of arm rings (or armlets as the tenth-century Vikings called them; they were arm circlets to the Romans) was associated with the wearing of bangles on the ankles, a holdover from ancient times. Bracelets have been worn in primitive and civilized cultures. The most ancient bracelets, like other jewelry articles of the past, come from graves and burial places of people whose daily lives were filled with trials beyond our comprehension.

A bracelet can be cast, plaited of twisted metal wire, or made of a number of single links joined together. Bracelets can be made of precious or base metal and gem-set (fig. 4-36). They can be a closed ring or opened on one side with a hinged clasp (fig. 4-37). There is a distinction made between a bracelet, which is flexible, and a bangle, which is rigid.

In the Middle Ages, the bare arm was not displayed and bracelets temporarily went out of style. However, in the Renaissance, both men and women fancied bracelets. An inventory of Henry VIII's jewelry lists seventeen bracelets, including one that was set with diamonds, rubies, and pearls. In northern Europe, the gold bracelet was most frequently found in the form of a chain worn over the sleeve. In the warmer climates of the south, gem-set or pearl-set bracelets were worn on bare arms.

During the bracelet's peak of popularity, it was set with gemstones, cameos, medallions, and pearls and decorated with enamel. But with the lace cuff and ruffled sleeve of the seventeenth century, arm jewelry was of little use. Occasionally, the fashion-conscious woman would wear a pair of pearl bracelets or a bracelet with enamel on one side and precious stones on the reverse side.

Since nothing exerts such tyrannical force as fashion, bracelets, or circlets, again adorned the wrist at the end of the eighteenth century. This time, however, the rage was to wear up to three at one time on one arm. The cameo was the dominant gemstone. Multicolored gold, filigree, and granulation were popular. The Victorian-style mesh bracelet was reproduced in the 1940s and can be mistaken for the antique. Both were of large scale; however, the reproductions are usually stamped whereas the Victorian bracelets were seldom stamped. Bangle bracelets of the late Victorian period may also be confused with later reproductions but can be separated by the karat stamping. The scale of jewelry often leads to dating certainty, an element present in some diamond bracelets. From the 1920s until the present, women's diamond bracelets have been popular. The straight-line diamond bracelet of the 1930s has been recycled as a lighter and smaller version called the tennis bracelet in the 1980s. Art Deco bracelet reproductions are common in the marketplace, especially the wide, geometric-style diamond bracelet. Frequently, this contains both diamonds and colored gemstones, often caliber-cut rubies and sapphires. When appraising this, remember that a genuine Art Deco bracelet with diamonds may double in estimated value if the piece also contains genuine colored gemstones.

4-36. This bracelet has good design, balance, and harmony. According to its designer, Norbert Brinkhaus of Canada, it has textured 18K yellow gold and two carats of diamonds. (*Photograph courtesy of Diamond Information Center*).

4-37. A distinction is made between a flexible bracelet and a rigid bangle, also between a bangle with or without a clasp.

4–38. When questions about markings arise, such as the identity of the one shown here, look in the *Jewelers' Circular Keystone Brand Name and Trademark Guide* and *Canadian Jeweller's Trademark Index* for help. A helpful book of American manufacturers is *American Jewelry Manufacturers* by Dorothy Rainwater.

4–39. Some jewelry is marked with stock numbers; others have patent numbers.

Questions and Answers About Bracelet Appraisals

Q. What is this (fig. 4-38) bracelet mark?
A. From Saudi Arabia.

Q. This clasp (fig. 4-39) has patent numbers and other maker's marks. How do I interpret this?
A. The 14K is the karatage of the gold. POM in a triangle is the mark of the manufacturer Pomerantz Jewelry Company, Inc. The patent number is only a very rough guide to the age of a piece. This patent number is on the clasp, so we assume it is for the clasp's design. You can learn about patents by researching the "Official Gazette of the U.S. Patent Office," available at major libraries. Or you can order a copy of a patent from the Commissioner of Patents and Trademarks, U.S. Patent and Trademark Office, Washington, D.C. 20231. Patent number 1 was assigned about 1836. The numbers were larger than 550,000 in 1896; larger than 1,500,000 in 1926; 2,300,000 in 1946; 4,000,000 in 1976. This information may be of greater import in establishing antique jewelry age than for modern or estate goods.

Caution! Only a patent number will say *Pat.* or *Patent.* The other numbers you may find on jewelry without the prefix *Pat.* or *Patent may* be the makers' stock numbers.

Q. I have a difficult time counting melee in some diamond bracelets and in describing on a report which diamonds fluoresce and which do not.
A. There is a simple solution if you have access to a copy machine. Place the bracelet (if it will lie flat) on the light table and get a photo reproduction of it. Then use that photo reproduction to point out the diamonds in a piece that fluoresce, those that are chipped, and the stones with various quality grades. You can do this by drawing an arrow to the appropriate diamonds. A picture copy of the bracelet is also an easy way to count melee (fig. 4-40).

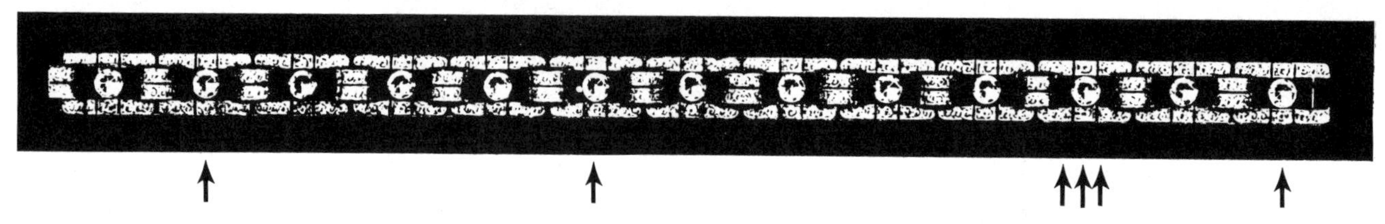

4–40. A photocopy of a piece of jewelry, made with the item lying flat on the light table, is an excellent way to count melee and indicate fluorescence in gemstones. Arrows indicate diamonds with soft blue fluorescence. *(Courtesy of Tom R. Paradise)*

MODEL APPRAISAL NARRATIVE

Writing a Description of a Bracelet

4–41. Gold bracelet.

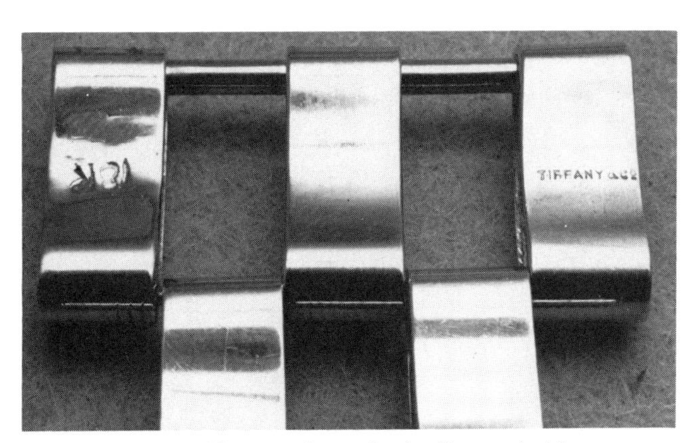

4–42. Clasp on bracelet in figure 4–41.

One (1) lady's gold bracelet, stamped and tested as 18 karat.

The bracelet is of die-struck manufacture and is constructed of a series of rectangular open boxes with ribbed design front and polished smooth backs. The boxes are stacked by twos and threes in fourteen (14) alternating rows. A long hinge pin placed vertically at the side of each row holds the stacks together and gives flexibility to the bracelet.

The bracelet is 8 inches long, including the clasp, 2¼ inches wide. The closure is two fold-over clasps that embrace the hinged pin in open spaces between two boxes.

The total weight of the bracelet is 16 dwts.

The bracelet is marked Tiffany & Co. and is stamped with an 18K mark.

Current Replacement Value $________________

____

Appraiser

The Brooch

The brooch, once called *fibula,* was a functional piece of jewelry used by nearly all European peoples from the time of the Bronze Age. The brooch, a Mediterranean invention, developed from a simple safety-pin-type object (fig. 4-43), designed specifically to hold a garment together, into an item of body adornment.

The *spina,* described by Webster's as "a thorn or spine," evolved into a pin used by early Romans and was made of metal. This became a clothes fastener called *spina ferrea* before it was known as a *fibula* and finally as a brooch.

There must have been a great and universal demand for the early brooch because this useful article is found buried with the remains of many warriors, chiefs, and their wives. In *Antique Jewelry,* Burgess tells us that the *fibula* fastened the Roman toga with sleeves secured from the shoulder to the wrist. One side of the toga lay at rest on the left shoulder, the other fell over the arm.

Early Scandinavian women wore embossed disk-shaped garment fasteners (fig. 4-44) of precious and base metals chained together in pairs. The influence of the design is felt to this day in Scandinavia where solid silver pendants in flat and circular shapes incised with symbolic and exotic motifs (fig. 4-45) are still favored.

Following the cloak clasp of the twelfth and thirteenth centuries, a hook-and-loop neckline fastening called an *agraffe* was substituted for the brooch. These ornamental pieces were fashioned with garlands, swags, rosettes, or leaves. The agraffe was important in the jewelry of the Gothic period, and its characteristics reflected both the region where the goldsmith lived and worked and the taste of the wearer. During the sixteenth century, the fibula, cloak clasp, and agraffe were important at various times before the brooch reappeared in the seventeenth century.

The seventeenth-century brooch was usually decorated with enamelled backs and gem-set fronts. The ladies of the court of the Sun King, Louis XIV, were known worldwide for their colorful brooches that served not just as functional items to secure the opening of a garment but also as decorations that called attention to a woman's graceful bare neck or plunging décolleté.

Diamond-set brooches became the fashion during the eighteenth century. The motifs were small sprays of flowers, branches, garlands, and wreaths. The pattern books of jewelry designs of the middle 1700s show large, open bow brooches of a type known as *Sevigné* (fig. 4-46). This was a large, floppy type of bow

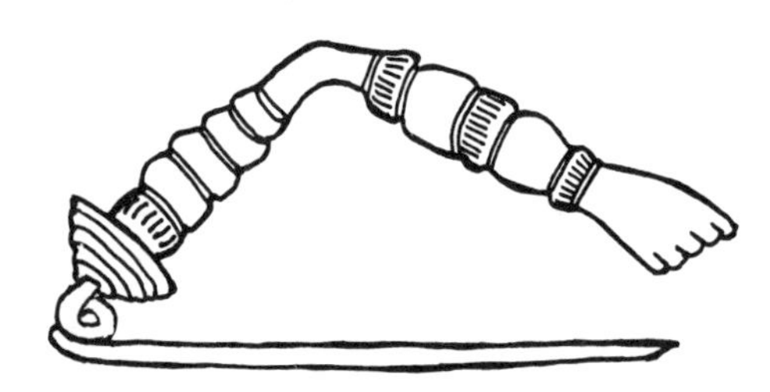

4–43. An early *fibula,* the forerunner of the brooch. (*Illustration by Elizabeth Hutchinson*)

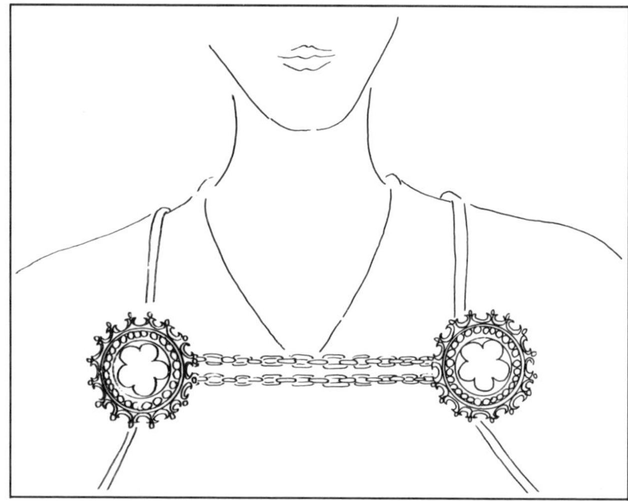

4–44. Disk-shaped garment fasteners were worn chained and in pairs. (*Illustration by Elizabeth Hutchinson*)

4–45. Traditional Scandinavian motifs in solid silver and pewter are very popular. The same motifs and designs used in early Scandinavian art of the tenth century are still in use today.

4–46. The Sevigné bow was a droopy-style bow of the eighteenth century. *(Illustration by Elizabeth Hutchinson)*

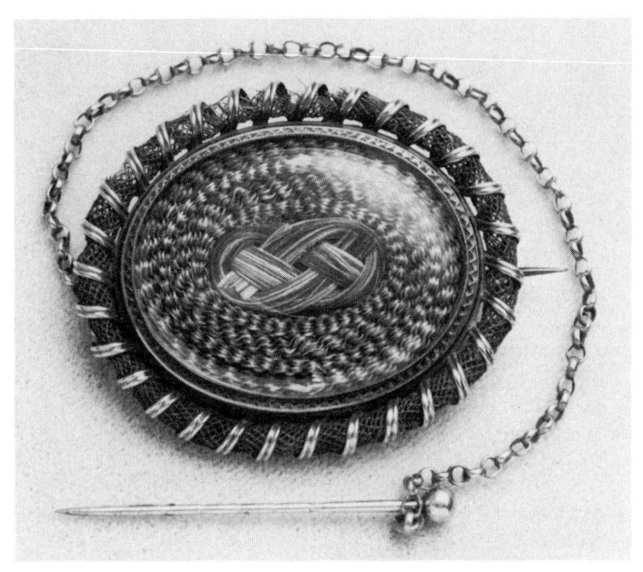

4–47. A memorial hair brooch with gold fittings.

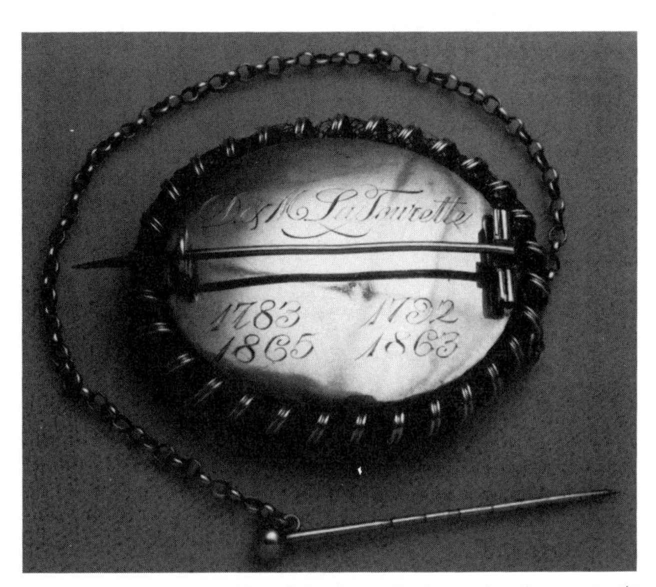

4–48. The reverse side of the brooch shown in Figure 4–47 inscribed with apparent birth and death dates.

design. It is also interpreted to mean a bow made of gold or silver in an open-work pattern set with numerous diamonds. In the early stages of the Sevigné bow, the style was simple, but later this style became more naturalistic with unequal bow loops and asymmetrical dangling ends. The difference this makes to the appraiser is in a proper narrative and design terminology and awareness of probable age.

At the end of the eighteenth century, the brooch was often used as a frame for a miniature portrait. The brooch frame was oval or round and usually set with pearls or other precious gemstones plus rich detailing in gold. This was succeeded by the simple frame cameo-set brooch or a Wedgwood porcelain brooch.

During the 1830s Gothic brooches, after a Neo-Renaissance style of scrollwork and enamel, were the fashion rage and a greatly preferred style. Jewelers created works in naturalistic motifs like flowers and butterflies, and this remained in vogue for the next fifty years.

Romanticism was a great influence on the evolution of brooch styles. The use of human hair—one's own or the hair of a loved one—was preserved under glass in a brooch or locket. Frequently, hair of a couple was preserved together (figs. 4-47 and 4-48). Far from *always* being a mourning or memorial piece, this type of jewelry also celebrated love, remembrance, and friendship. An entire industry made hair jewelry in the early nineteenth century.

Although hair brooches were frequently worn, there is evidence that some women preferred keeping the hair out of sight in lockets fitted in the back of jewelry. These receptacles were known as *boxes* in the nineteenth century and were widespread as jewelry fittings. The front of the piece often gives no indication of the presence of a box behind, but the depth of these boxes will indicate to the appraiser that they were intended for hair and not for photographs. By 1875, hair jewelry was completely out of style and almost nothing will be found that can be dated beyond that time.

Clasps are often clues to dating. In the early nineteenth century, pins were longer than the body of the brooch. When fastened, the pin point was worked back into the fabric acting as a safety device.

Multiples of small scatter pins were worn by women in the 1940s, only to be replaced by the bold and significant single brooch of the 1950s. This large style

4–49. New designs based on old motifs. (*Photograph courtesy of Diamond Information Center*)

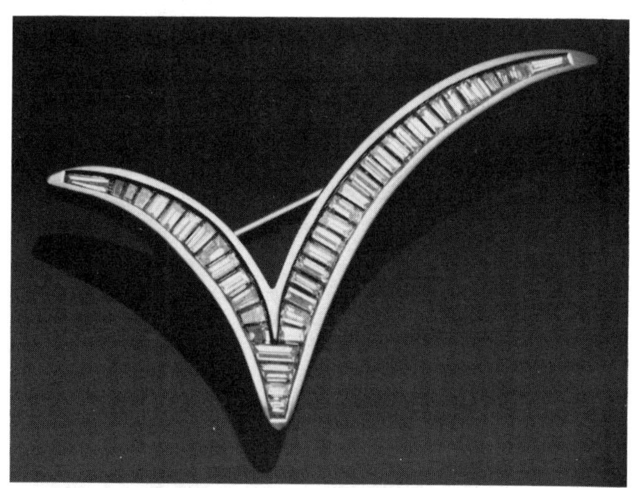

4–50. A 1980s' interpretation of a swag motif. (*Photograph courtesy of Diamond Information Center*)

4–51. Geometric styling in the 1980s. (*Photograph courtesy of Diamond Information Center*)

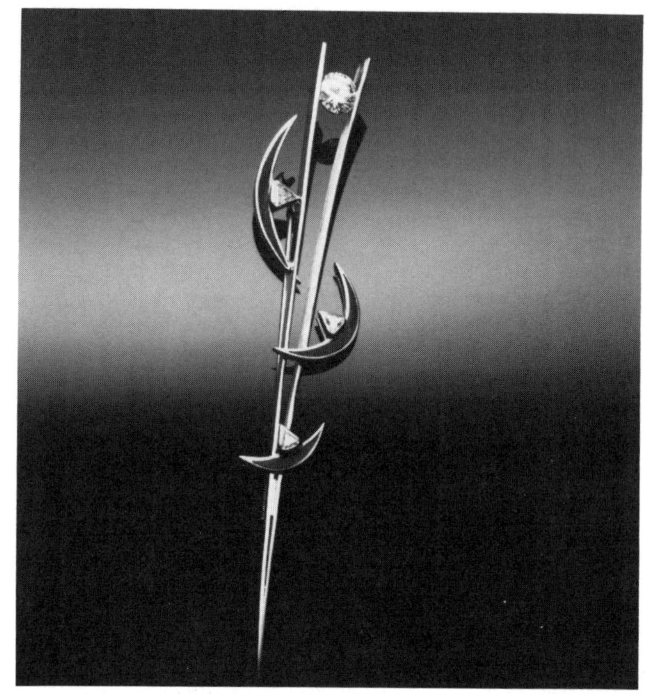

4–52. The crescent motif updated. (*Photograph courtesy of Diamond Information Center*)

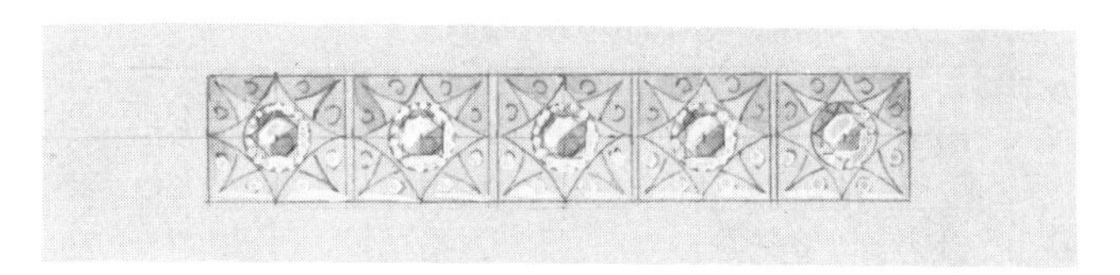

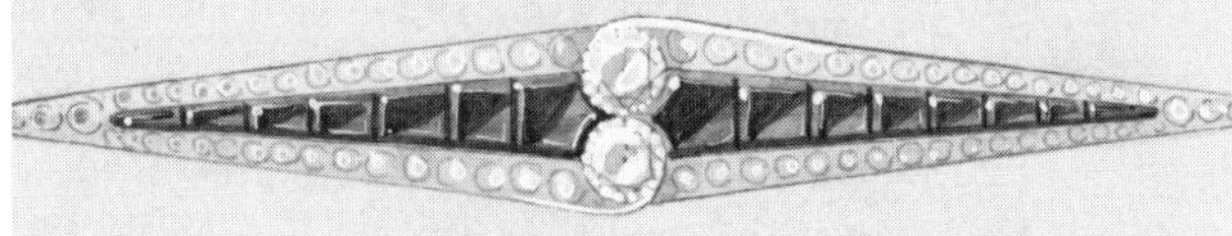

4–53. Jewelry designs done in 1925 from the Goldbaum jewelry firm, Pforzheim, West Germany. (*Courtesy of Tom R. Paradise, T. R. Paradise & Co.*)

1950s' brooch has been revived in the 1980s, but with a sleek, contemporary look and muted-metal finishes with smaller gemstones (figs. 4-49 to 4-52). Close inspection of the new works will show that present-day designers have not strayed far from classical styles (fig. 4-53) but still draw inspiration from earlier motifs, such as geometrics, flowers, swags, crescent moons, and circles.

Recognizing Motifs and Periods

Learning to recognize motifs and classify specific periods is a process of slowly accumulating information. In studying the following illustrations of brooches, note the specific characteristics that make them examples of their individual periods.

Revival jewelry (figs. 4-54 and 4-55) is part of many appraisals. The brooch pictured is of the nineteenth century Renaissance revival. It is in girandolé style and set with emeralds, cultured pearls, and table-cut and rose-cut diamonds. Observe the overall design of the brooch, the cut of the gems, the shape of the pearls, and the metal. Study the piece for any signs of marks or hallmarks. Look for evidence of repairs. Note the clasp. Note the settings for the gemstones.

Here is what we learn from observation. The brooch is in a typical Rococo swag-and-garland motif. The majority of diamonds are table-cut, suggesting those not table-cut are replacements. The round emeralds are all full-cut brilliants, except for the rectangular-cut emerald set in the center of the brooch. Emeralds have smooth polished tables without any abrasions, scratches, or wear (a positive indication of new or replaced gemstones). The pearls are drilled through in the centers, all the same size and shape, and show no wear (a strong suggestion that the pearls may be cultured and replacements).

The metal tested as silver over 18 karat gold. The piece shows signs of enamelling. There are no hallmarks, but the style suggests German or French.

A significant number of repairs are evident and lead soldering is readily apparent. Numerous metal tears and breaks around the scrollwork still exist.

The clasp is a tube catch of the type popular about 1890; however, it looks as though it may have been

4–54. Revival-style brooch.

4–55. Reverse side of brooch shown in figure 4–54.

4–56. Edwardian brooch of high quality.

4–57. Close-up detail of brooch in figure 4–56 showing millgrained heads and knife-edged mounting details.

added to the brooch, or be a replacement. The article appears handcrafted except for the round die-struck heads soldered onto the mounting. These heads look like a later vintage along with the rectangular head in the center of the brooch. All diamonds are in *closed-back settings.*

Conclusion: This article is probably a marriage of three different periods. The table-cut diamonds and the way they are set, along with the motif of the design, suggest a piece of jewelry from the late eighteenth century. Clasp, emeralds, and cultured pearls all seem to be part of later periods. If you have read the jewelry correctly, the message will be clear that at sometime in the past vanity has created a hybrid. It also contributed to a lowered value.

Figures 4-56 and 4-57 show a brooch from an entirely different period. Observe the style, the cut of the diamonds, the metal, and the embellishments to the gemstones. This is what close observation tells us. The round brooch is in a foliate motif, typical of the Edwardian period. The piece is set with old European-cut diamonds and a few old mine-cut diamonds in typical (for the period) millgrained heads. The brooch is set with one full-cut round brilliant diamond, an obvious replacement.

The brooch is of platinum, an essential metal in determining the Edwardian style and period. Knife-edge settings and lacelike designs are typical of the 1900–1914 era. Also during this period, pearls (natural and cultured) became popular mixed with diamonds. A small amount of cultured pearls began to arrive on the jewelry scene in the first decade of the twentieth century. The cultured pearl novelty turned into a steady stream of imports by the 1920s.

Figures 4-58 and 4-59 are pictures of a very well-crafted brooch from the late 1950s. Observe these elements: quality of manufacture, condition, motif, cut of the gemstones, metal, and care of finish.

This is what the brooch reveals about itself. First, the design is a combination of Art Deco, Retro, and Contemporary, and best described as transitional. The

4–58. Aquamarine and diamond brooch, circa 1950.

4–59. Reverse of aquamarine brooch in figure 4–58.

motif is geometric with modified arrow or pendulum design. The contemporary look produced in platinum and decorated with diamonds and large colored gemstones is typical of the late 1950s.

The brooch is of die-struck manufacture with light hand assembly. It is set with fine quality full-cut round brilliant diamonds and fine, even colored, well-matched genuine aquamarines. The aquamarines are set in heads better than the average quality, indicated by the extra protection in "V" tips around the ends of the stones.

The back of the brooch has been well finished. The brooch has double pin stems with two (2) clasps and each clasp has a safety catch.

The fine-quality gemstones, platinum material, care of finish, and quality of findings reflects an expensive item of jewelry. Its excellent condition speaks of the care given by the owner. The conclusion is that this is a fine quality brooch, circa 1950–1960, of sizable retail replacement value. If one is seeking fair market value on a similar brooch, the most appropriate market would be auction sales.

MODEL APPRAISAL NARRATIVE

Writing a Description of a Brooch

4-60. Gold brooch.

4-61. Reverse of brooch in figure 4-60.

One (1) yellow gold brooch, stamped and tested as 14 karat.

The brooch is of cast manufacture, in an openwork design, and in a butterfly motif. The yellow gold is in a bright finish.

The brooch measures 1½″ h. x 2¼″ w. (36mm x 57mm).

The brooch has a 14 karat gold pin stem and pin clasp with a safety lock on the clasp.

Total gold weight of the brooch is 7 dwts.

There is no manufacturer's mark.

Current Replacement Value $__________

____

Appraiser

Cameos

Cameos are among those jewels currently enjoying a new wave of popularity. Although used in all jewelry items and in many objets d'art, a cameo is most distinguished used as a gem in a brooch. True cameos have multilayers of carved figures in relief, but this art form belongs more to the past than today. Modern cameos are mostly carved in shell and show only one color.

Cameo and *intaglio* are two important words to the jewelry appraiser. The most obvious difference between the two is that a cameo is cut in relief and the design is fashioned so that it stands out above its background. The cameo has no functional role and was made solely for its beauty and ornamentation. An intaglio is cut so that the design is below the surface of the background; the seals of the ancients are its ancestors.

The most popular shape for a cameo is the oval, although other shapes have been used such as rounds, squares, rectangles, pear-shapes, and others. Cameos range in size from very small for earring use to the size of a dinner plate. The large ones are a special domain to collectors.

Twentieth-century cameos have lost much of their artistry with the development of such technical involvements in the carver's craft as machines and ultrasonic engraving. Sadly, cameo art of the last fifty years seems to be stuck in the quagmire of mediocrity.

Hardstones and shell are the materials for cameos, but for general wear the shell cameos are the most sought after. Hardstone cameos can be quite heavy, especially when set in a large gold frame, and not as wearable on modern lightweight garments. Shell cameos in rings, on the other hand, are at risk. The owner of a cameo ring may find that after a few years the face or figures carved on the cameo have worn away. This can devalue it by 50 percent.

In 1805 Napoleon fanned the desire for the cameo by creating a prominent and artistic level of excellence for gemstone engravers when he opened the French Academy for Gem Engraving.

As a collectible, cameos enjoyed a popularity during a neo-Classical revival from the late eighteenth century until about 1840. At that time, a scandal hit the industry when collectors learned that fake antique cameos were flooding the market. That information took the passion from the collectors and interest in the jewels dimmed.

The situation reversed again with the rise of the middle class, which began to travel more. The archaeological digs of Pompeii were great tourist attractions and people naturally wanted a memento of their visit. Cameos, in a great variety of materials, were being carved especially for the visitors. Lava cameos from Mt. Vesuvius at the Pompeii site were attractive in colors of matte gray, olive green, brown, beige, mustard yellow, and black. In fact, lava cameos are so common in this period that an old lava cameo being examined and evaluated as in the Romantic-Gothic style can be tentatively dated as Victorian by the appraiser. Cameos of various colors set as a series in bracelets or necklaces are not uncommon. The cameo will often be carved with an archaeological motif or with mythological creatures such as satyrs or bacchantes, like the bracelet in figure 4-62. Other important cameo materials were lapis lazuli, ivory, coral, and jet. Shell cameos are typically Victorian; the shell craft revived in Italy in 1805.

4–62. Lava cameo bracelet. (*From the collection of Dr. and Mrs. Edgar B. Smith*)

In the late-nineteenth century, cameos with diamonds, pearls, and precious stones were the rage as a fashion statement on a Grecian-inspired highwaisted dress. A stylish lady wore cameos on her belt, necklace, bracelet, diadem, and on a headband. Between 1870 and 1880 there was considerable carving of cameos. The result was that pieces became larger and larger, with crossed details and mythological scenes such as

4–63. Nymphs on cameos were stylish in the nineteenth century.

4–64. Mid-nineteenth-century cupid cameo suite of jewelry.

the three Graces, chariots with horses, and warriors charging skyward. Nymphs (figs. 4-63, 4-64, and 4-65) with flowing robes and swirling garlands, as well as winged creatures and Cupids, were in demand. Cameos went out of fashion when the desire for miniature portraits in brooches became the rage. In addition, the demise of the shawl, a perfect accessory for a cameo brooch, did much to reduce the desire for this jewel.

Even though cameos have worn well over the centuries, faces have sometimes been lost, a nose has broken off, or rubbing has eroded details of costume and figure. Costumes and faces can often help one date a cameo, according to Seattle antique jewelry dealer Karen Lorene. Lorene has presented her observations on antique and modern cameos for discussion in seminars and lectures. She makes this point: Look at the profiles, especially noses, because they have undergone dramatic changes over the years. The classical Greek and Roman faces with straight noses (figs. 4-66,

4–65. A rose gold and agate cameo suite of jewelry, circa 1850. The mounting has applied yellow gold leaves. The subject is a dancing nymph and cupid.

4–66. A finely carved cameo with good detail and Roman nose.

4–67. The straight Roman nose helps date the cameo.

4–68. The cameo lady took on a jowly look by mid-nineteenth century. *(From the collection of Dr. and Mrs. Edgar B. Smith)*

4–69. Costuming and subject matter are clues to circa dating cameos. This cameo with the muse Terpsichore is Victorian.

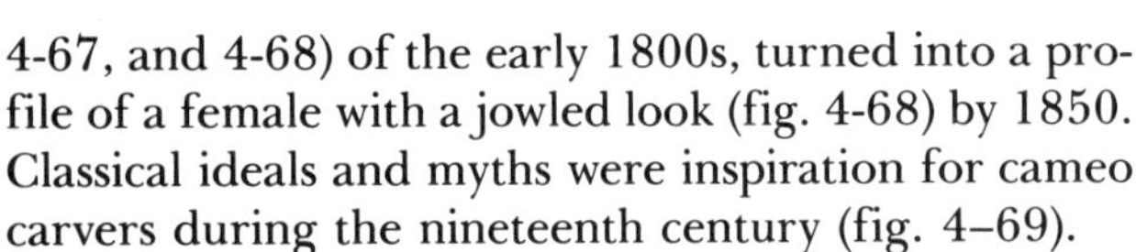

4-67, and 4-68) of the early 1800s, turned into a profile of a female with a jowled look (fig. 4-68) by 1850. Classical ideals and myths were inspiration for cameo carvers during the nineteenth century (fig. 4–69).

The profiles of two great actresses, Sarah Bernhardt and Eleonora Duse, were popular subjects from about 1917 to 1930. Many of the cameos were carved with the actresses wearing costumes from their most celebrated roles. By 1980, however, the cameo profile has changed into a cover girl look with a decidedly upturned nose and a great mane of flowing hair. Cameos (fig. 4-70) are modern Italian works reflecting the contemporary woman as seen by twentieth-century carvers. All are shell and range in size from 10mm x 15mm to 36mm x 50mm. They are priced at wholesale from $100 to $800.

Some of the value of a cameo is in its age. To establish age can be confusing and difficult, requiring study and observation of profiles, facial features, body proportions, subject matter, and attention to those details inherent in ancient cameos.

When appraising cameos, examine hardstone material to determine whether it is dyed, enhanced, natural, or a composite; be alert for molded glass, assembled glass, and molded plastic.

Check cameos for cracks. Cracks will reduce the value of a cameo with the amount of value reduction aligned to the nature and predictability of the crack. Most cracks found in shell cameos are vertical and can be faint to severe. To observe the cameo, hold it up to the light. Faint lines indicate cracks. There is no repair or restoration. Cracks in hardstone cameos, if severe, may reduce value by half or more, depending upon whether or not the crack impairs the subject matter.

Study the carving. The deeper the carving and the more layers used in the design, the more valuable the cameo. The cameo in figure 4-71 (p. 88) is an example of a deep carving. Hard stones like tourmaline and corundum are more valuable if the quality of the carving equals the quality of the material. One clue to identifying a machine-made cameo is that machine-carved has

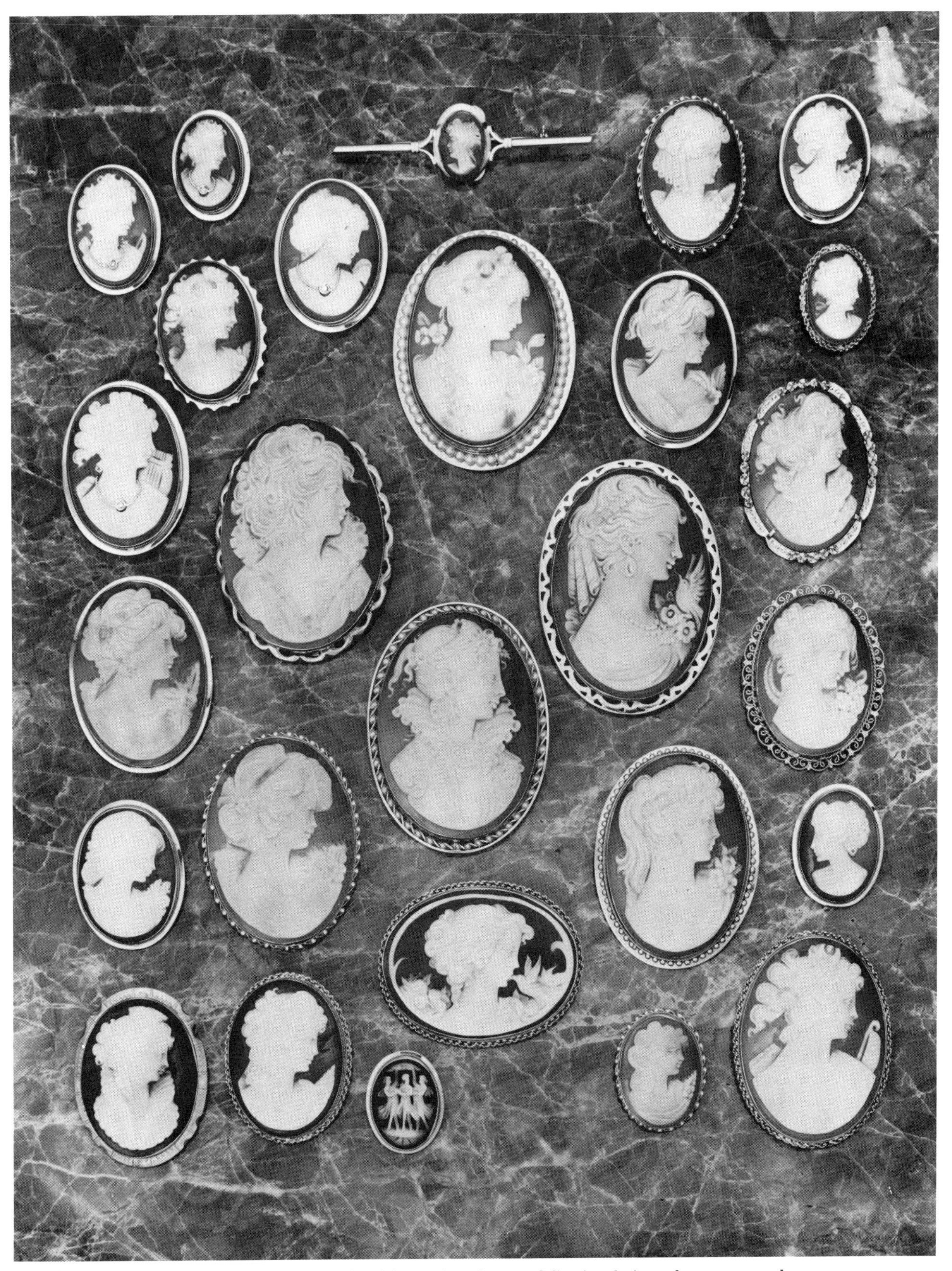

4–70. Modern cameos look like cover girls with an abundance of flowing hair and an upturned nose. (*Photograph courtesy of Pat deRobertis, Italian Connection*)

4–71. An example of a deeply carved cameo. This one is gray and white chalcedony.

precise lines sharp to the touch; handcarved usually has more fluid lines; undercutting and tool marks are sometimes apparent. Study the fittings (figs. 4-72 and 4-73) for added value.

In the final analysis, a cameo must be evaluated as an art form and the appraiser should have scholarship in carving material and subject and be able to pass judgment on composition, proportion of design, finish, and detail. Signatures may be found but may prove difficult to authenticate. Nevertheless, any signature found on the cameo should be noted on the appraisal document. It should be detailed as to exactly where on the cameo it was found.

In cameo appraising, knowledge of comparative quality is critical. Nothing is more valuable to appraisers than their own study and actual observations and examinations. Take every opportunity to study cameos in private collections, museums, auctions, and gem shows. This is an area where there is almost no academic instruction available. The best teacher is personal interest, lively curiosity, and cautious validation.

4–72. This cameo pendant becomes a brooch when the bail and its support is removed. The upper portion is removed from the bail finial and the tubular casing slides off. The cameo is dated circa 1870. (*Photograph courtesy of Waller Antiques, Inc.*)

4–73. Reverse of the pendant/brooch cameo shown in figure 4–72. A unique fitting such as this adds value. (*Photograph courtesy of Waller Antiques, Inc.*)

4–74. The subject of this cameo carving is the classical Greek tale of Leda and the swan. Modern interpretation by German carvers in hardstone.

Subject Matter

The knowledge of Roman, Greek, and Christian mythologies can be very helpful. Many late-eighteenth-century neoclassical revival cameos have scenes that look strange, especially the Greek myths (fig. 4-74). Understanding the subject matter will enable the practitioner to circa date more conclusively and validly.

Some classical Greek figures, pagan gods and goddesses, and mythological figures that are frequently seen on cameos and depicted in a variety of forms are the following:

Abundance	Wheat ears and inverted cornucopia
Acratus	Bacchus as a winged Pan
Adonis	Seen wearing a hunter's costume, usually with Venus, also with a dead boar and a dog
Aeolus	Guides Bacchus to Ariadne; he is a bearded, winged figure.
Aphrodite	Goddess of love and beauty. Also known as Venus. The myrtle is her tree; the dove her bird—sometimes, also, the swan.
Apollo	The master musician with a golden lyre. As the Apollo Conservator, he is seated; as Apollo Sol, the deity, he has a radiated head.
April	A youth dancing in front of a statue of Venus
Atlas	Bearded nude figure seated on a mountain
Concordia	Holds an olive branch and stands between two standards
Castor and Pollux	They (The Dioscuri) wear oval helmets and are mounted on horses. They usually have a star as their symbol.
Demeter	Goddess of the harvest, sister of Zeus. Her attribute is a stalk of wheat or grain.
Dionysus	Youngest of the gods; also called Bacchus, god of wine
Endymion	Asleep in the arms of Morpheus; sometimes seen with Diana and preceded by Love holding a torch.
Hebe	Goddess of youth, she sometimes appears as cupbearer to the gods.
Hermes	Messenger of the gods, also known as Mercury. He is known by his winged sandals and wings on his hat.
Fortune	Seen with the sun and crescent; two cornucopia; one foot on prow of a ship; and as Fortuna Manes holding the bridle of a horse.
Eros	God of Love (Cupid in Latin)
Janus	The two headed deity, looks backward and forward
Justice	Seen with scales and sword
Leda	Shown with the swan, which is Zeus in disguise. Zeus came to Leda, wife of the king of Sparta, in the guise of a swan. From this union were born Castor and Pollux, the twins, and Helen, who became the cause of the Trojan war.
Muses	The nine muses are draped and can be distinguished from nymphs by their long tunics and draped busts. Their distinguishable symbols are: Urania has a sphere at her feet, Polyhymnia has a roll and robe below her girdle, Thalia wears a mask and carries a pastoral crook, Terpsichore plays the lyre, Calliope wears a cloak around her waist, Clio carrys two thongs and chastises one of the muses, Melpomene carries a dagger and wears a mask, Erato writes love poetry, Euterpe recites lyric poetry.
Naiads	The water nymphs. They dwelt in brooks, springs, and fountains.
Narcissus	Stands at a fountain
Peace	A goddess carrying the infant Pluto

Piety	Veiled and holding a cornucopia
Poseidon	Neptune, Lord and Ruler of the Sea. The trident is his attribute.
Providence	Female figure leaning upon a column with a cornucopia
Satyr	Usually a creature with horns and the face and legs of a goat.
The Three Graces	Splendor (Aglaia), Mirth (Euphrosyne), and Good Cheer (Thalia)
Triton	The trumpeter of the sea, his trumpet was a great shell.
Zeus	Lord of the sky. The thunderbolt is his symbol; his bird, the eagle.

Cameo Markets

The great danger in valuing cameos is from a lack of awareness of the subtleties of this art. The average appraiser, seeing only a few pieces of this type of jewelry a year, may become complacent about quality and values. One way to keep current with pricing is to become closely associated with the collector's market. The collector's market is a type of third-tier market (collectors buying and selling among themselves) that is alive and vital, even when there seems to be no other sales information on more conventional levels such as auctions.

A recommended dealer/collector in cameos is George C. Houston, 550 South Hill St., Suite 1478, Los Angeles, CA 90013. Houston has been a collector since 1938 when he began an apprenticeship as a precious stone cutter. He has collected hundreds of engraved gems such as stone and agate cameos from Germany and shell cameos from Italy. Much of his collection dates back to the 1920s and in some cases before 1900. His current catalog collection lists one hundred seventy cameos with comments about their execution, size, coloration, condition, and so on. The catalog is a wonderful resource for the appraiser because Houston also lists prices. The range of materials and subjects is broad, from a light brown lava cameo of the 1800s with a cherub motif for $300 to a red and white carnelian collector's prize signed "Pauly" for $800.

For studying collections of antique cameos, the Hull Grundy collection in the British Museum has a famous cameo group, and the Walters Art Gallery in Baltimore, MD, has some outstanding examples of early cameos.

Questions and Answers About Brooches and Cameos

Q. What is the proper nomenclature to describe the direction a cameo portrait faces?
A. If the figure faces right, *dextral*; if left, *sinistral.*

Q. Is there a quick-sight identification to distinguish silver-on-gold from platinum-on-gold jewelry?
A. David Atlas of D. Atlas & Co. handles hundreds of antique items yearly and offers this observation for instant assessment but warns it may not be 100 percent accurate: If the white metal on the mounting is seen as the majority of color, it is probably silver; however, if the white metal is seen as a thin layer with the yellow color as majority, the white is probably platinum. Atlas adds that silver also oxidizes and may look dark while platinum remains bright. He advises making acid tests several times to develop a comfort level with the accuracy of this sight technique.

Q. Is a Victorian Scottish pebble brooch set in gold much more valuable than one set in silver?
A. Yes. The gold Scottish cairngorm brooch would be about four times more valuable than a silver one.

Q. Antique jewelry expert Joyce Jonas of New York University speaks about the "spirit of design" as a help in appraising jewelry. Exactly what does she mean by that?
A. Jonas believes that many jewelers tend to be a bit myopic when it comes to dating jewelry or being able to appreciate the overall design and construction. "One of the elements of value," she says, "is to be able to recognize if a piece reflects the life, energy, and spirit of the period . . . the spirit of design." Edwardian jewelry is a direct reflection of the elegance of the era. French jewelry was lighter in weight and more delicate in scale than English Edwardian jewelry, keeping to the "spirit of design."

Q. What is *memento mori* jewelry?
A. *Memento mori* means in memory of the dead and refers to jewelry given to family and friends of the deceased at funerals, usually metal-encased, sepia on ivory brooches that show a sobbing figure, tomb, weeping willow trees, angels, or flowers. Often locks of the deceased's hair were used. The fashion for this kind of jewelry continued until the late eighteenth century when people could no longer afford such tributes. The brooches may be valued today between $400 and $1,000 depending upon condition, materials, and provenance.

MODEL APPRAISAL NARRATIVE

Writing a Description of a Cameo

4–75. Cameo brooch.

4–76. Reverse of the cameo brooch in figure 4–75.

One (1) genuine shell cameo set in a hand-fabricated mounting stamped and tested as 10 karat yellow gold.

The brown and white shell cameo measures 25mm x 35mm and is a classic woman's profile, circa 1840. The piece is artfully carved with fair relief. There are no cracks or other damage to the shell cameo. There is no signature. The cameo is bezel-set in the mounting.

The mounting is handcrafted and has stamped and applied leaf decoration. Greek key-fret motif is cut out around the bezeled cameo. The edges of the mounting have been rolled backwards upon the mounting for ornamental effect. The mounting measures 2″ h. x 1¾″ w x ½″ d. (50mm x 43mm x 13mm). The pin and stem are base metal and are replacements. There is a new safety catch. The cameo mounting shows some signs of repairs with lead solder.

The mounting weighs 7 dwts., including shell cameo.

Current Replacement Value $________________

____

Appraiser

Beads

In the beginning there were beads. Long before humans had knowledge of the inherent beauty in a rough diamond or colored gemstone, beads satisfied the yearning for body decoration. Beads predate rock paintings, carved bone figures, and all other forms of recorded art, according to bead authority Peter Francis, Jr., at the Center for Bead Research, Lake Placid, N.Y. In a 1988 *Bead Report* article, Francis writes that only one other decorative medium is older than beads: body painting.

Stone as a bead material came into use in the Neolithic period. Excavation sites of many former civilization have yielded bead finds. Shell, ivory, and the teeth of carnivores have been widely used for decoration, social statement, and magico-religious beliefs.

In past societies, everybody, regardless of status, owned beads. While royalty was entitled to pearls, peasants had to be satisfied with pot-clay beads. Beads are as popular today as ever before and are imported into this country by the ton from Europe, the Far East, and West Africa. Due to technological advances, beads are now more perfectly formed than they have been in the past. Simple or elaborate, beads in necklaces, brooches, or earrings satisfy tactile and visual senses with a variety of textures, designs, and materials.

The most popular beads of all time are pearls. The 1980s' fashion favorites include lapis lazuli, malachite, rose quartz, amber, coral, gold, silver, jade, and multicolored tourmaline. In research on beads, Pansy Kraus, former editor of *Lapidary Journal,* writes that early humans used beads not only around the neck but also around the torso, arm, and legs. The ancient Egyptians wore cylinder-shaped beads in collars, while their body girdles were made of multiple strands of small beads strung with bone spacers. Glazed faience, a material of ground quartz fused by use of an alkali and covered with a glaze of finely pulverized colored quartz, was a favorite Egyptian bead material. Others were carnelian, agate, and steatite (soapstone).

In East Asia and India, where only the wealthy could afford fine jewelry, the poor had to make beads from whatever material was handy. In most cases this was various colored agates and jaspers. Trade beads of hardstones and glass were introduced by merchants following the caravan routes. Ship captains used beads for barter as well as money.

Within the last decade, beads of all types and materials have come increasingly into vogue. The current fashion trend is for beads strung with extra large accent beads, or objets d'art such as netsukes or ojime beads. This mixing of everything and anything is meant to enhance and individualize necklaces to suit the mood of the wearer as well as dramatically accessorize the costume.

4–77. The six strands of beads photographed are of several varieties, all of exceptional quality. (*Photograph by Tino Hammid, courtesy of House of Onyx*)

When appraising bead necklaces that have carved jade, gemstone pendants, or netsukes, take the current price of the accent item into consideration and do not include it as part of the strand of beads. The accent should be judged separately, much as a clasp merits individual attention, and that figure should be added to the total sum of the beads. Since the bead field is vast and expanding in popularity among the public and jewelry designers, appraisers need current pricing information. For in-depth reports about beads, the most helpful resource is the magazine *Ornament,* P.O. Box 35029, Los Angeles, CA 90035. Consistent and timely articles about new finds, trends, styles, and designs of beads fill every issue. Information is also available from The Bead Society, P.O. Box 2513, Culver City, CA 90231 and The Bead Society of Greater Washington, P.O. Box 70036, Washington, DC 20088. The Bead Museum, a nonprofit incorporated museum in Prescott, AZ, has information to share about beads from ancient, ethnic, and contemporary cultures.

For help with 1989 bead values, the following price list from the House of Onyx, Lexington, KY, may be a helpful guide. Prices may differ in your area because of regional preferences and supply and demand. Research your local market!

4–78. Bead price lists are essential support data for the appraiser. Beads have always been popular consumer items. *(Photograph courtesy of House of Onyx)*

Pictured from top to bottom (fig. 4-77):

A 16-inch-long, six strand twisted necklace of baroque and round cultured black pearls with 14K gold beads and clasp, $1,050.

A necklace of carved lapis lazuli, 37 inches long, with 18mm accent beads and 10mm round lapis beads, 14K gold beads and gold clasp, $3,095.

An eight strand twisted coral necklace, 28 inches long with cultured pearls and a 14K gold clasp, $775.

A 8mm bead malachite and 10mm bead amethyst necklace, 24 inches long and accented with 14K gold beads and clasp, $225.

Persian turquoise beads graduated from 10mm to 14mm, 25 inches long, with a yellow-color silver clasp; the turquoise weighs 97.33 grams, $500.

A 24-inch-long necklace of 12mm black onyx and white howlite with goldtone clasp, $75.

The necklaces (fig. 4-78) are all 16 inches long, except for one strand, and most have a gold tone clasp. The beads are well matched, all center-drilled, all of fine quality material with excellent polish. Priced from top to bottom:

- Rhodonite, 12mm beads, $35.
- Snowflake obsidian, 10mm beads, $20.
- Lavender jadeite, graduated from 8.5mm to 11mm, 22-inch-long strand with 14K gold clasp, $450.
- Tiger eye, 10mm beads, $25.
- Carved ivory, 12mm beads, 22-inch-long necklace with ivory clasp, $75.
- Hematite, 10mm beads, $20.
- Blue lace agate, 10mm beads, $20.
- Sodalite, 12mm beads, $30.
- Epidote, 10mm beads, $25.
- Rose quartz, 10mm beads, $25.
- Malachite, 10mm beads, $60.
- Rutilated quartz, 10mm beads, $30.

MODEL APPRAISAL NARRATIVE

Writing a Description of a Bead Necklace

4–79. Lapis lazuli and gold bead necklace.

One (1) single strand necklace containing sixty-one (61) round lapis lazuli beads measuring an average of 7.5mm in diameter, and sixty (60) 14 karat yellow gold spacer beads measuring an average of 3mm each alternating with the lapis lazuli beads.

Lapis lazuli beads are opaque with even, dark blue color with a medium polish. Pyrite and quartz inclusions are seen in the beads. There is no indication the beads have been color-enhanced. Beads are evenly center-drilled.

The necklace measures 24 inches in length, including the clasp, and has a 14 karat yellow gold fishhook filigree clasp, stamped as 14K.

There are no knots between the beads.

Current Replacement Value $________________

____

Appraiser

Beads are measured in millimeters ($^{1}/_{25}$ of an inch) and are usually sold in temporarily strung 16-inch lengths. Table 4-1 may help determine the number of beads on a strand. The number given is standard for most materials:

Table 4-1. Approximate Number of Round Beads on 16″ Strand

Size	Pearl or Bead
3mm	128
4mm	100
5mm	80
6mm	66
7mm	57
8mm	50
9mm	45
10mm	40
11mm	36
12mm	33
14mm	29
16mm	25
18mm	23
20mm	20

Common modern bead shapes: round, disk, bicone, fluted, melon, square, carved, barrel, tubular, and navette. Note the following points about beads on the appraisal document:

1. Type and quality of material
2. Bead shape
3. Quality grade and condition
4. Bead sizes in millimeters
5. Quality of drilling
6. Use of spacers
7. Length of strand
8. Clasp

Questions and Answers About Beads

Q. What are mummy beads?

A. They are ancient Egyptian beads mostly made of faience, a ceramic substance of soapstone and powdered quartz, with glaze made from copper compounds. The men who made faience were called baba (firer of glaze). The beads were made on an axis, probably string or thread, with the string coated with a paste compound, rolled into cylinders and scored into sections of different lengths. They were then dried, coated with glaze, and fired. Huge quantities of the beads have been found in Egypt. They were made into single and multiple necklaces and also sewed onto garments, belts, aprons, and sandals.

Q. What is an ojime bead?

A. A Japanese ivory bead that is often confused with a netsuke. The ojime bead was a type of fastener for a personal carrying case. In the early years of the Edo period (1615–1868), ojime beads were simple spheres, but they developed into miniature sculptures and exhibited intricately carved designs. They are sought after today as focal points in bead necklaces and are valued as individual items.

Q. Beads are often strung on materials other than silk or nylon thread. What is tigertail and foxtail?

A. Tigertail wire is one of the quickest methods of stringing. It is nylon fused on wire. Foxtail is a flexible woven metal chain generally used on heavy, large-holed or sharp-edged beads. There are four sizes of foxtail that are used in bead stringing, and they range from fine to very heavyweight.

Q. What is the French wire method used to finish a strand of beads or pearls?

A. Appraisers should be aware and able to distinguish a well-finished strand of pearls or beads. The French wire method requires more skill than normally employed in stringing. French wire is fine wire wound into a flexible springlike tube and used at each end of a necklace to form a proper loop for joining the clasp ring to the strand. When used, this technique signifies a more professional job and a stronger closure.

Q. When noting measurements of a necklace, is the clasp included?

A. No hard and fast rules apply because the clasp must often be individually appraised. Make certain that it is clearly stated on the appraisal if the overall measurements include the clasp or not.

Ethnic Art Jewelry

Stamping, impressing, chip carving, and *fire gilding* are some of the terms you should be acquainted with when appraising in the field of ethnic art jewelry, which is largely beads. All the aforementioned techniques are used in the decorative process.

Stamping is an old method of basic stamping to repeat design; impressing is the direct stamping of symbolic designs onto the face of a finished object; chip carving is a relief pattern carved in metal with a small chisel; and fire gilding refers to a technique used on nongold metals to make them look like gold. You will find these methods and more used on beads, artifacts, and amulets—items of various ethnic heritage—combined into articles of jewelry.

Not yet accepted in the fine jewelry category nor in costume jewelry, ethnic jewelry pieces are not usually found in the showcases of antique jewelry dealers, although many of the components are antiques.

4–80. Ethnic art jewelry. This piece by Judith Ubick retails for $985. It is an antique ivory netsuke on a jade ring with an ivory fan holder, jade and brass buttons, jade and serpentine beads. All components are of Chinese origin. *(Photograph courtesy of Judith Ubick)*

4–81. This lovely antique sterling silver, gold, and coral necklace is highlighted with an antique prayer box from Tibet. Designed by Judith Ubick, the value is $1,200. *(Photograph courtesy of Judith Ubick)*

Appraisers unfamiliar with this kind of jewelry (fig. 4-80) may classify it in a lower category with a resulting undervaluation. This is especially true if they take a cost approach and value the metal and gemstones for their intrinsic worth only.

Ethnic jewelry has gained momentum in the past decade as a popular and exotic style of ornamentation. Its proponents are designers who create one-of-a-kind necklaces, bracelets, and anklets, and the consumers who view it as a unique personal statement in jewelry.

One of the best known designers in this field is Californian Judith Ubick, 15226 Friends, Pacific Palisades, CA 90272. She has been stringing necklaces and incorporating important artifacts and amulets into them as a central theme for the past seventeen years. By her own count, she produces some 2,400 original pieces a year. The necklaces range in retail price from $200 to $5,000. They are not signed, but they do have a distinctive design signature appearance. The 36-inch-long antique silver bead, antique silver, coral, and gold prayer-box pendant necklace (fig. 4-81) is strung on hand-crocheted wire, typical of Ubick's work.

How will the average jeweler be able to tell if the piece is matchless or just another Hong Kong import? The *average jeweler* probably cannot tell. But the appraiser must be able to make the differentiation between the unique strand of beads and the ordinary, because ethnic art jewelry can often be quite costly.

Ubick believes that an appraisal of her jewelry must be prepared by one who is not a stranger to the original use and antiquity of the materials. She also insists that it is critical that the appraiser know the market from both the buyer's and seller's position. Ubick stresses that some of the components in her jewelry are no longer available at any price, and so the originals skyrocket in value. For example, she points out that China has closed the door on export of antique merchandise, so "No antique jade, antique silver, Tibetan turquoise, Mandarin necklaces, or antique carved stones are coming out of the country." China has firmly shut off the supply of genuine artifacts.

How can the appraiser work efficiently in this market? The ethnic art jewelry market is separate from antique jewelry in America and most European countries, and very little is seen at antique jewelry shows. Neither is it being sold with any regularity at auctions; thus the appraiser must rely on other resources such as designers themselves, periodicals, and books about the subject.

Publications

Ornament magazine is the bible of the bead fancier and therefore the appraiser's choice for information, prices, and leads on additional sources. Some good book resources are: *The History of Beads,* by Lois Sherr Dubin; *Ethnic Jewelry,* by John Mack; *Ethnic Jewelry,* by Dona Z. Meilach; *Kayamanan,* by Ramon N. Villegas; *Out of The Past: The Istanbul Grand Bazaar,* by Burton Y. Berry; *Ivory,* published by Harry Abrams, Inc.

Pearls

Pearls, the oldest gems known to civilization, are classics of jewelry tradition. Regrettably, they are the least understood by appraisers. To gain finite knowledge of pearls, one needs total immersion in historic pearl fact and fiction, market cycles, forecasts, demographic buying patterns, market demands, and consumer psychology. All of these elements are important in the accurate value estimation of a strand of pearls.

This jewel from the sea has a romantic history glorified in verse and song, with mention in the pages of the Bible as well as in the writings of the Greek poet Homer. Further, articles on the eloquence and desirability of the pearl regularly appear in modern jewelry trade press and consumer publications. Little wonder then that the abundance of publicity surrounding the pearl confuses the issue of value for both client and appraiser. To get an adequate understanding, the appraiser is required to do focused research and employ deductive reasoning.

Pearl History

For centuries it was believed that the pearl was not only a symbol of purity and innocence but also a powerful antidote for poison as well as treatment for a variety of physical ailments. The writings of Confucius (552–479 B.C.) mention freshwater pearls as valuable gifts for over a thousand years. During the same period, they were also believed to be a treatment for diseases of the eye and ear. In the thirteenth century, the *Lapidaries of Alfonso X of Castile* reported that the pearl was an excellent cure for heart disease, sadness, timidity, and melancholia. To this day in Persian Gulf countries, pearl powder is used for indigestion, while in India crushed pearls treat lung and eye disease, headaches, and smallpox. In Japan, crushed pearls are used as diuretics, according to pearl expert Toshio Ishida, and also ground for use in curing dyspepsia. Powdered pearls are used in health foods as a source of calcium and to cure the common cold!

In the book *Pearls* by Shohei Shirai, the oldest pearl ornament still in existence today is recorded as a single-strand necklace that was unearthed during an archaeological dig at a Persian king's palace in western Iran. No mention is made of where this necklace is kept, but Iran is inferred.

Pearl detractors are few; in fact, throughout history men and women have enjoyed wearing pearls. In the sixteenth century, Henry VIII of England had pearls embroidered onto his hats, mantles, and shoes, but Elizabeth I had even greater use for pearls. She is often seen in portraits wearing a pearl collar, pearl girdle, pearl carcanet, and a gown entirely embroidered in pearls.

As the pearls of the Orient flooded Europe, they became essential jewelry for royal ladies. In the court of Charles I of France, they wore deep décolletage that provided an ideal frame for ropes of large pearls. In portraits of "grande dames" and royalty of the seventeenth century, pearls are seen in hair decorations, collars, necklets, brooches, girdles, and earrings.

Pearls have been popular in the United States since the discovery of freshwater pearls delighted the early colonists. Indigenous to the lakes and rivers of North America, the freshwater pearl mollusk was also a good food source. Many reports exist of early-day "pearl rushes" at the turn of the twentieth century to rivers in Ohio, Wisconsin, Tennessee, and Arkansas. A strong mussel-shell and pearl industry flourished until the early 1920s. Several factors combined to curtail the industry: plastic buttons came onto the market eroding the market for mother-of-pearl buttons, and the stock market crash of 1929 triggered a collapse of pearl prices and the start of the Depression. The industry has never regained its former vitality.

Soon after World War II, the cultured pearl industry in Japan began to concentrate upon building a foreign following. With the introduction of the cultured pearl, demand was not tied to supply as it was before 1900 but became the partner of fashion. The Japanese effort to bring cultured pearls to the United States has been very successful. The U.S. market has become one of the strongest worldwide. The Japanese Pearl Exports Association points out a record $180 million dollars of pearls were exported to the United States in 1984, and the figure rises annually. U.S. pearl wholesalers claim that the Japanese are cultivating more of

4–82. Moshe Kaufman says pearl color grading is best with a white background, not gray. (*Photograph courtesy of L. Kaufman & Sons*)

4-83. The pearl companies sort pearls for luster and color using white backgrounds. (*Photograph courtesy of The Cultured Pearl Association of America and Japan*)

the larger pearls (from 6.5mm up); therefore smaller pearls (5.0–5.5mm) are scarce. For the appraiser, that kind of information mandates a careful research of the market to select true comparables and obtain current prices.

An example of how a fashion trend is established on a national scale is found in Barbara Bush, wife of President George H. Bush. She is seldom photographed without her multistrand 8.0–8.5mm cultured pearls or her necklace of imitation pearls. Considering her importance, one can conclude that many American women will follow suit, and the popularity of larger-size cultured pearls should go up.

The Japanese Cultured Pearl

Current statistics show that more than 95 percent of all pearls sold in the world today are cultured. There are three types of cultured pearls known: the Akoya (Japanese) cultured pearl, the South Sea pearl, and the freshwater pearl, often called the Biwa pearl.

The Akoya cultured-pearl process begins when the oyster is about three inches in diameter. The oyster is opened a few centimeters, and a nucleus and sliver of mantle is inserted. The mantle is taken from a live oyster; the nucleus is a round mother-of-pearl bead made from the Pigtoe clam of the Mississippi drainage area. The implanted mantle tissue will envelop the bead and that will trigger layers of nacre around the nucleus by the host mollusk. To make a good pearl, hundreds of layers of nacre are needed with a normal growing time of three to four years. Many factors affect the growth of a pearl: weather, temperature or salinity of the water, predators, and pollution. Those pearls that make it to harvest and processing will be sorted by size, color, luster, shape, quality of the skin, and cleanliness of the nacre.

How the Experts Evaluate Pearls

Color

Moshe Kaufman of L. Kaufman & Sons, Inc., a New York firm of pearl experts, discounts what the Gemological Institute of America teaches as pearl color grading background: "I have never met one pearl dealer who uses the color gray, as GIA instructs, as a background to grade color. Never! How can you accurately see color using a gray background? Every pearl dealer has a white pad on his desk (figs. 4-82 and 4-83) and uses it to grade the color of pearls!"

Color evaluation of pearls is controversial as a value factor. The Gemological Institute of America uses and teaches color as an element in their approach to quality grading. Instructors say that in weighing the qualities of two pearl strands identical in all aspects except color, a cream color strand with greenish overtone may be less desirable to American women than white pearls with a pinkish overtone. This may be true in certain parts of the world. U.S. importers know that fair-complexioned Americans, Canadians, and most northern Europeans like white- or cream-body color pearls with a slightly pink or rosé overtone. Japanese women prefer white pearls with greenish overtones, while Latin women prefer white pearls with golden overtones. So while appraisers appraise for a specific client and region, they should be aware of subtleties of market preference. How color impacts value is a more difficult and subjective call than grading color in gemstones, even for the most notable pearl connoisseurs. The message appraisers should heed is that while color is significant, it should not be granted *over*importance in the final estimation of value. All markets do not dictate the same color pearl! There are six main body colors found in pearls: white, cream, pink, golden, black, and green. Pearls are also found in blended color varieties such as lavender, silver, and gray.

Lazar Kaufman, patriarch of the Kaufman company, believes that Japanese cultured saltwater pearls are commonly treated to improve color: "Most dealers in the United States will not tell you this," he said, "but at least 99 percent if not all Japanese saltwater cultured pearls are dyed or bleached to even out or lighten body color. Bleaching is not a detectable treatment." Kaufman noted that, if the pearls are Japanese black cultured pearls, they are most certainly dyed. "There are no natural black cultured pearls from Japan;" he says, "the pearls that are black, brown, or peacock colors usually have a yellowish body color to start with."

Color grading pearls can be confusing. Because of different light sources, even some experts have been fooled. Human color vision is not just a reaction to

4–84. Grading pearls by natural north daylight in Kobe, Japan, Kaufman evaluates his selection. (*Photography courtesy of L. Kaufman & Sons*)

wavelengths of light but also a response to an object's apparent color change depending upon its surroundings. This phenomenon explains why pearls look different on different skin tones. It is a subject with its own war stories. In the Kaufman offices in New York City, pearls are graded by natural north daylight because light and reflection make a difference in color perception. This was vividly demonstrated when Kaufman was inspecting some hanks of pearls in Japan in front of an open window. "The examination showed the pearls to be of acceptable color and they were bought and carried back to the United States. Then when the boxes were opened here in the office, the pearls looked awful. They were off-color, brownish not bright. The problem, as we later found out, was that when the pearls were being examined in front of the open window, a building totally painted red stood across the courtyard and it caused a reflection upon the pearls."

Most gemologists understand the light principle as it is applied to colored gemstones, but many fail to recognize this same concept also works for color in pearls. As with colored gemstones, pearls are ideally graded in natural north daylight (fig. 4-84) before ten o'clock in the morning. If natural light is unavailable, fluorescent light is the next best lighting.

Even in the actual Japanese pearl market auction room, light for examining and grading is closely controlled. The room itself is so constructed that wherever one sits the light is always the same. There is no direct sunlight or artificial light. Each table is covered with a cloth of intense whiteness on which the pearls are placed to determine their luster and color.

Cleanliness

The most important criterion for assessing the value of a pearl is the perfection of the skin, that is, lack of blemishes. Blemishes, or spotting, are those wrinkles, cracks, gaps, bumps, or dull spots on the surface skin of pearls. One imperfection is called "orange peel," and it does resemble the skin of an orange. For pearls with heavy spotting, values plunge. Some pearls with only a spot or two may have one of those spots eliminated by using the spot as the point for a drill hole.

Judging cleanliness is a matter of experience in handling various grades of pearls. Moshe Kaufman has this to say about the correct way to examine pearl strands: "An experienced pearl dealer will pick up the strand, hold it in his hand, and inspect the strand *from the bottom up,* not from the top down! The best pearls are usually blended into the center of a strand, so look at the ends—the way old-timers judge—because if you find spotted pearls in the end section, you will surely find the rest of the strand has blemished pearls."

Spotted or blemished pearls can be peeled, that is, the outer layer can be peeled away like the layers of an onion. "This is a time-consuming and exact process," says Kaufman, "that is only accomplished by an expert." A few people in New York claim expertise in this field. They refuse to give any information about the process, preferring to keep their secrets. A few years ago, Graduate Gemologist Kim Hurlbert wrote from Los Angeles about a laboratory incident in which she believed she stumbled across the secret of pearl peeling. She was working on a ring that had a 7-mm pearl. She accidently put the ring in a diluted solution of sulfuric acid. "Evidently," she said, "the bowl of water I was using to rinse the pearl ring in had been previously used to rinse pickling solution from other pieces of jewelry. The acid in the water caused the surface of the

4–85. A brownish material often found on freshwater nucleated pearls can be peeled away, but the process is time-consuming.

pearl to become frosty and completely dull. Polishing with a soft bristle wheel followed by tripoli and rouge on muslin wheels removed what had become a powdery layer, and this action returned the pearl to its original lustrous appearance!" Laying further claim to the discovery of the peeling secret, Hurlbert reported that an industry expert had long hinted at the necessary removal of a powdery layer during the peeling process.

Another cosmetic artifice akin to pearl peeling is used occasionally on nucleated freshwater pearls. Pearl dealer Douglas Sparrow of San Francisco acknowledges there is a process but insists it is so time-consuming that only a few pearls are treated in this manner. "When nucleated freshwater pearls are fine in every respect except for a brown stain (fig. 4-85) that is sometimes found on the surface skin, they can be made salable by scraping or peeling." Sparrow demonstrated with an X-acto knife on the pearl shown in figure 4-86, where the top and one side was covered with a brownish scum-like material. On the left side the pearl is shown with the brown stain; on the right the stain has been scraped away and the luster restored. "I have seen pearl scraping with a rubber wheel," Sparrow said, "but it burns up the pearl if the wheel is too hot." He mentioned superfine sandpaper or ruby polish as two other possible treatments.

Another cosmetic trick Sparrow spoke of is the use of a gluelike mixture to fill holes and give color. "This is a totally unacceptable method of treatment and must be disclosed as a treatment of the material," he said. As to the scraping remedy, presently less than 10 percent of nucleated pearls get this kind of cosmetic treatment. Sparrow believes it could become more prevalent if the present poor water and nutrient environmental conditions continue.

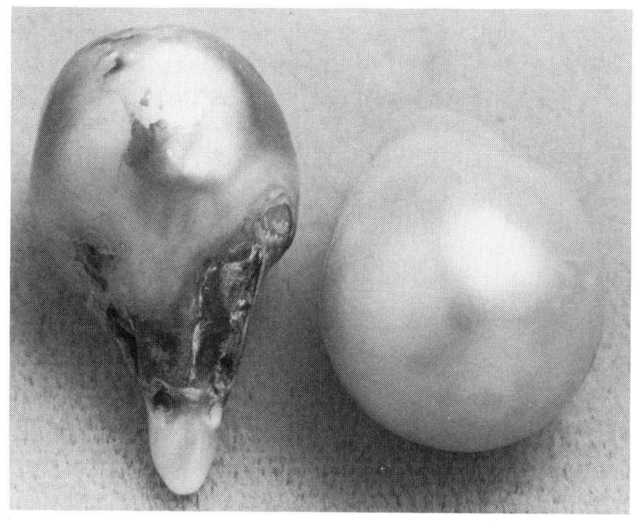

4-86. Unless the brown spots are scraped away, the pearl is relatively valueless.

Cultivation

Nacre thickness increases beauty and durability of pearls with the ideal thickness being 1mm. A three-year growing period has been normal in Japan, but is now decreasing. The desired size of the pearls and the financial strength of the cultivator to endure waiting periods have determined the years of growth allowed the oyster. Small pearls from small nuclei in small oysters have been harvested after only one season's growth, May to January. Now a new process patented in Japan may allow even shorter growing times. The Matsushita Group has developed and tested a substance called protein cell coating. This new agent allows a harvest of round cultured pearls after only a *six-month growth period.* The formula is based on the fact that a calcium crystalline structure on the surface of the nuclei is regarded as a foreign substance by the oyster. When the nuclei are coated with an organic protein matter, the mollusk demonstrates a lower level of rejection. The protein cell coating serves as a physiological activator causing layers of nacre to develop at a much faster rate around the nucleus. Interestingly, the developers of this method found that by using the protein cell compound around the bead implant, together with the traditional mantle tissue, the survival rate is increased to 79.4 percent. The old survival rate without using the coating was 60.3 percent. Tests using 6-mm nuclei-coated beads were conducted in Japan in 1987 and proved highly successful as far as yield. Accordingly, in 1988, Matsushita manufactured 1,125 kilograms of the protein cell compound for distribution to Japanese pearl farmers.

What does all this imply for the appraiser? In effect, a new era is about to begin with cultured pearls on the market having a thinner nacre, probably masked by fine color. Appraisers *must* become familiar with the various elements that make up the value of a pearl and keep up with all the technological advances in pearl cultivation. It would be useful to establish a running dialogue with a friendly pearl importer.

An innovative way to look at a pearl's cultivation was explained by Moshe Kaufman: "When holding a strand of pearls, spin or twirl them around at eye level and see if there is a change of colors, that is, dark and light. Pearls with thin or uneven nacre display dark and light spots. You can almost see the bead sometimes because the nacre is so thin. A pearl with a heavier nacre coating will not show as much of a tonal change."

Luster

The more mirrorlike the surface reflection of a pearl, the better the quality. A pearl's nacre can sometimes be determined by the luster, since the thicker the layer, the finer the luster.

Luster is an important factor in determining quality. Every pearl has a bright spot, a highlight, where light is reflected. The area around this highlight will have a glow that seems to emanate from under the surface. Occasionally a pearl will show an iridescent play of color called the *orient.* The orient is more commonly seen on baroque-shaped pearls.

Shape

The most valuable pearls are round. In a well-matched strand (fig. 4-87), the pearls are similar in shape and give the viewer the overall impression of evenness. Half-round pearls are those with flat bottoms. They are also called button pearls. Irregular pearls are termed *baroques* (fig. 4-88). A strand of baroque pearls should also give the impression of evenness, although the individual pearls are not round-shaped. Some baroques have part of the nacre ground away to improve the shape, but the improvement makes little difference in the final estimate of the pearls' value. In freshwater pearl necklaces, pearls should be matched for shape, although they will seldom be round.

4–88. The nomenclature for different pearl shapes. *From left:* round, off-round, baroque or drop, dyed tadpole shape. The off-round pearls on the end (*right*) show orange peeling.

Recognizing Cultured Pearls

There is no visible difference between natural and cultured pearls, so sight identification by the appraiser is impossible. X ray is the only method that will show differences in structure (fig. 4-89). Pearls that are to be X-rayed for identification can be loose or strung.

Under ultraviolet light, cultured pearls often have milky-white luminescence and under X rays a green one. Appraisers have been told they can look down the drill hole with a loupe and see the bead nucleus surrounded by the dark conchiolin ring for certain knowledge that the pearl is cultured. This is a cumbersome and unsatisfactory exercise. Another test, helpful but not conclusive, is specific gravity. In the case of most cultured pearls, the specific gravity is greater than 2.73 while that of natural pearls is often lower.

4–87. Stringing matched pearls. A well-matched strand will be given higher value consideration than one of mixed shapes. (***Photograph courtesy of The Cultured Pearl Association of America and Japan***)

Valuing Japanese Saltwater Cultured Pearls

The only way to have positive assurance of valuation for a particular string of pearls is to do personal research in the pearl market. Pearls vary so greatly in size (fig 4-90), color, and other qualities that the appraiser is urged to visit several pearl dealers to look for exact comparables before finalizing value. Although grading and pricing systems are invaluable resources, no price guide offers figures as supportable as personal research estimates. Nevertheless, price lists are basic material, and Kaufman & Sons has developed a grading and pricing system. They offer it as a helpful guide to pricing Japanese cultured pearls. Prices recorded were in effect at the time of this writing, January 1989, but appraisers are reminded that price changes occur swiftly and unevenly in the pearl market. Also, it should be noted that the prices are adapted to the Kaufman Company's own pearl-grading system.

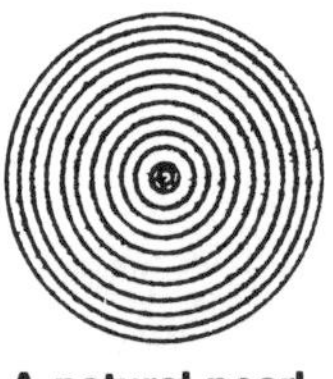

4–89. Structure of a natural and cultured pearl.

The price guide in Table 4-2 is based on the cultured pearl quality elements: color, cleanliness, cultivation, and luster. The color used is a white body color with pinkish or rosé overtone, the most popular pearl color in the United States. The prices are wholesale, based on 16-inch-long temporarily strung pearls that are well-matched and spherical.

The following Table 4-2 lists *wholesale* prices.
A represents a pearl strand 80 percent perfect.
B represents a pearl strand 60 percent perfect.
C represents a pearl strand 40 percent perfect.

Table 4-2. Japanese Saltwater Cultured Pearls—Wholesale Price*

Pearl Size	A	B	C
4½ x 5	$175	$100	$70
5 x 5½	$215	$125	$90
5½ x 6	$280	$160	$110
6 x 6½	$345	$200	$140
6½ x 7	$448	$260	$180
7 x 7½	$724	$420	$295
7½ x 8	$1,100	$640	$448
8 x 8½	$2,100	$1,200	$840
8½ x 9	$4,140	$2,400	$1,680
9 x 9½	$9,300	$5,400	$3,800
9½ x 10	$16,000	$9,500	$6,000

*Current prices January 1989. *Compiled by L. Kaufman & Sons, Inc., New York.*

South Sea Pearls

What is a South Sea pearl? The term usually defines a pearl over 11mm in diameter that may be natural or cultured. The cultured South Sea pearl undergoes the same cultivation processes as the Japanese Akoya pearl, except a different type of mollusk is used. The white-lip oyster is used south of the equator; the gold-lip and black-lip oysters are used mostly north of the equator. The type of mollusk used influences the color of the pearls. The most desirable shape is the perfect sphere; the most prized color is white with a pinkish overtone and the most valued is the black. The giant black-lip pearl oysters were used for trial cultivation in 1912 in Okinawa but didn't catch world attention until 1978 when a single-strand necklace of cultured Tahitian black pearls commanded a $500,000 sales price in New York. Black pearls are cultured in the waters around Australia, Indonesia, the Philippines, Thailand, and Burma. These warm waters produce oysters that can harvest 15mm- to 16mm-size pearls.

Even though the existence of black cultured South Sea pearls has been known by professionals for many years, gem quality black pearls (pearls perfectly spherical, solid black, and flawless as in figure 4-91) are limited in supply. South Sea cultured black pearls may be classified into three color groups: brown-black, green-black, and black. Natural South Sea black pearls do exist but they are extremely rare and seldom seen outside of museum or royalty collections.

Some cultured South Sea pearls are dyed black, and the appraiser should be aware of this treatment. The use of X-radiography may help detect it by showing the pattern that is typical of silver nitrate dye. Because metallic silver deposited from the silver nitrate

4–90. Sizing of cultured pearls. (*Photograph courtesy of The Cultured Pearl Association of America and Japan*)

4–91. Two beautiful South Sea cultured pearl necklaces. The Tahitian natural black South Sea cultured pearl necklace is $85,000 wholesale. The pearls are from 11mm at the clasp to 15.1mm at the center. The white South Sea cultured pearls are $165,000 wholesale and measure 11.6mm to 14.8mm. The center pear-shape pearl drop is 17.2mm × 14mm. (*Photograph courtesy of Albert Asher South Sea Pearl Company*)

Table 4-3. Cultured South Sea Pearls—
Wholesale Price Per Piece (per Pearl*)

Size in mm	Price by Classification				
	A	B	C	D	E
10 to 10.9	$2,500	$2,000	$1,400	$900	$500
11 to 11.9	$3,000	$2,300	$1,800	$1,200	$800
12 to 13	$4,500	$3,300	$2,700	$2,200	$1,500
13.1 to 14	$5,500	$4,300	$3,200	$2,600	$2,000
14.1 to 15	$7,000	$5,500	$4,000	$3,200	$2,600
15.1 to 15.9	$11,000	$8,500	$6,000	$4,000	$3,500
16 to 16.9	$20,000	$16,000	$13,000	$10,000	$7,000
17 and larger	$32,000	$26,000	$20,000	$13,000	$9,000

*Current prices January 1989. *(Courtesy of Albert Asher, Asher South Sea Pearl Co.)*

solution is opaque to X rays, it will show up as a white ring on the X ray. The conchiolin layer, which is transparent to X rays, will show up as a black ring.

Another method for detecting treatment is longwave ultraviolet light. Natural color cultured black pearls fluoresce a dull orange-red to red color when exposed to longwave ultraviolet, but dyed cultured black pearls do not fluoresce.

The Japanese Akoya blue pearl is often mistaken for a cultured black South Seas pearl. The blue cultured Akoya pearls are not South Seas pearls, however, and are not produced from the black-lip oyster. They have transparent bluish, green, or deep blue colors, and an overwhelming majority seen by appraisers have been dyed blue. In general, the cultured blue-dyed pearls will look suspect as the color of the dye is often garish and unnatural. Dye will sometimes be suspected because of the uneven luster of the pearl. Inspect the drill hole or the pearl string for signs of dye. Dyed cultured pearls are considerably less valuable than cultured natural blue pearls.

Albert Asher of the Asher South Sea Pearl Company in New York is one of this country's leading experts on natural and cultured South Sea pearls. To help appraisers understand the great value of these special pearls, Asher created a price list based on his company's quality grading system. "I have tried to categorize the South Sea pearls," Asher said, "but every pearl has its individual characteristics. For instance, at a recent pearl auction, I paid over $45,000 for a single pearl 17.6mm, and about $29,000 for a single pearl 16.4mm!"

The following prices are *wholesale* from the Asher South Sea Pearl Company. "Retail prices," Asher notes, "will vary from 50 to 90 percent markup over these wholesale prices, according to the jeweler."

The following is a key to the classifications in Table 4-3:

A Perfectly round, very clean, high luster, smooth skin. Color: cream rose, white with pink, light rose with very light cream undertone.

B All of **A's** specifications minus one of these traits only: luster better than medium or color silvery white, or pink with very light blue undertone.

C Round, nice luster, fine color, but one or two very tiny spots. Or as in **B** the color is either stark white or light, or light cream but bright; or luster is below medium.

D Color acceptable but either the skin is not very smooth, or somewhat "milky" as far as luster, or the pearl has some spots difficult to conceal.

E The pearl is slightly off-shape, or the color is too creamy, or with gray undertones, or pink with strong blue overpowering secondary hue.

Drop-shaped pearls and round button shapes with flats on one side will follow the **A, B, C, D** estimates providing the pearls are well proportioned. They are usually more desirable if they are in pairs than in single pearls. These shapes are excellent for earrings and are primarily used in that fashion.

4–92. A superb example of nearly flawless South Sea cultured pearls is this strand with pearls from 10.5mm to 15.5mm. Pear-shaped pearl earrings are 12mm × 15mm size. *(Photograph courtesy of Albert Asher South Sea Pearl Company)*

Baroque-shaped pearls command about 50 percent of the value given to round pearls. Pearls that have rough shapes or skins and those blemished or with poor cultivation (nacre) will be valued at about one third of the regular estimates.

South Sea pearl necklaces are usually sold in graduated pearl sizes. At present, the most popular is a necklace with cultured pearls from 10mm to 14.5mm. For a normal 16-inch-long necklace, measured without clasp, the strand will contain twenty-nine to thirty-three pearls (fig. 4-92).

The same classifications for single pearls and quality elements are applied to a strand. If too many of the negative features are combined, the price will fall sharply. For instance, a necklace of 10-mm to 14-mm cultured South Sea pearls that has excessive spotting, or is not well matched in color, will vary (wholesale) between $15,000 to $25,000.

A 16-inch-necklace using the quality elements cited will fall into these price ranges:

- Quality A – $250,000 to $300,000
- Quality B – $190,000 to $235,000
- Quality C – $140,000 to $175,000
- Quality D – $ 90,000 to $125,000
- Quality E – $ 40,000 to $ 60,000

In this category of evaluation, you must do personal research before making quality judgments.

American Natural Pearls

Little information is available on natural American pearls. What is known is that they were found in great quantity on the Pacific coast, concentrated around the southern end of Baja, California, especially around La Paz, in the nineteenth century. While the pearling industry on the American Atlantic coast has been exhausted, Mexico revived the pearl fisheries along the coast of Baja a few years ago. However, the supply of pearls from that area is extremely limited. Most of the pearls being used in Mexican jewelry at this time are still imported from Japan.

San Francisco gemologist Cortney Balzan has made a study of the American natural pearl. In so doing, he has amassed a considerable collection for research and for sale. A minute sampling of the collection is shown in figure 4-93. Balzan reports that his pearls are round to baroque and include many odd forms. They range in size from .05mm for a tiny seed pearl to approximately 8mm for a button-shape pearl. The colors vary from white, gray, golden, and purple to black. The luster is notably good and most are nearly blemish free. Balzan sells the pearls by the carat with a fair-quality one costing about $25 per carat wholesale. Congenial to sharing his research information, he can be contacted at the Balzan Gemological Laboratory, P.O. Box 6007, San Rafael, CA 94903.

4–93. A small selection from the collection of natural American saltwater and river pearls, assembled by Cortney Balzan.

Questions and Answers About Pearls and Mother-of-Pearl

Q. I have a box with small round pieces of mother-of-pearl (fig. 4-94) about the size of poker chips, engraved with elaborate designs. What were such chips used for?

A. These are chips and counters for games. Chips like the one pictured have been made since the time of the ancient Egyptians. The one illustrated was

4–94. A mother-of-pearl game counter.

probably made in the Orient in the eighteenth or early nineteenth century for export and sale in England. Some of the most ornate chips were engraved with a family coat of arms. Counters were made in many shapes. The shape defined the value of the chip for the game. Individual chips are seen often at gem shows and cost about $60 and up depending upon decoration.

Q. I have a pair of mother-of-pearl opera glasses to appraise. How old are they and what is their value?

A. Most of the pearl opera glasses seen in estate appraisals today were made in France in the latter part of the nineteenth century. Some of the earliest date from about 1860. Occasionally Sotheby's and Christies have these for auction. Average auction hammer prices are $450 to $600. The Asprey Company on Fifth Avenue in New York has a new pair of mother-of-pearl opera glasses in a fitted case listed in their catalog for under $450.

Q. What are Mikimoto pearls?

A. Japanese pearl farmer Kokichi Mikimoto received a patent on a method to culture pearls in 1908. By 1920 cultured pearls were being imported into the United States. The name Mikimoto equates with fine-quality cultured Japanese saltwater pearls.

Quality Recognition Techniques

The three pearl and gold rings pictured in figure 4-95 are typical of pearl rings seen in estate appraisals. Close observation of these rings will tell you about their quality, thus narrowing the markets that will be researched to estimate value. Observe the following: a quick inspection tells you these rings are not in a fine jewelry category but fall into the commercial and low-end retail market. The ring with the multicultured pearls has a stamped-out top that appears thin and has been assembled to the ring shank. The pearls are a mixture of sizes, colors, and qualities. This is discernible even in the photograph. Notice that some of the pearls are different shapes, baroque to egg shape. They are various colors. One is dull and lusterless (top right-hand corner). Several have medium luster while spotting and orange peeling is apparent. Most alarming of all is that they do not seem to fit into their settings. They are not pegged, so they must be held by prongs that seem to barely grasp them, and several prongs are missing. There is a 14K stamp on the shank, but it did not test as 14K.

4–95. In considering the quality grades of these three rings, it is evident that value estimations should not be made cursorily.

The conclusion is that the cultured pearls have been set in exchange for the original stones, probably colored faceted stones. Considering the stamped and assembled top, the brittle and missing prongs, the short length of the remaining prongs, and the Oriental style of the design leads to the conclusion that this ring was probably made in Thailand and styled in the Princess motif. This type of ring when new and set with garnets, dark blue sapphires, or black star sapphires, is common to Thailand. It can be purchased in 14K gold in Bangkok with some synthetic and some natural stones for $200 to $250 retail, natural stones in 18K up to $350.

The ring in the center has a mabe pearl in a cast mounting that is stamped as 14 karat. It has sixteen (16) cultured pearls set around the mabe. They all fit snugly in their settings and are of matched color, size, and luster. What can this ring tell us? Look at the obvious dark mark on the shank where it has been sized. The poor work, like any poor repair, lowers the value of the ring. Observe the mabe pearl. If one looks closely around the edges of the pearl, the mother-of-pearl bead that backs the mabe can be seen. The conclusion on this ring is that it, too, has had the center stone replaced. The mabe is too large for the ring and set in this manner, with short and stubby prongs, is in danger of being broken. Although difficult to see in the photograph, the prongs around the mabe pearl have been notched, obviously to hold a faceted gemstone.

The third ring is of cast manufacture, stamped as 14 karat and tested as 14 karat gold. The pearls look to be the same size and same quality. We can observe that the pearls, although baroque, have good luster. Spotting can be seen on the pearls. The pearls are obviously peg-set. The mounting also contains two small diamonds, and the condition of the ring is excellent. Conclusion: This estate ring is in almost new condition. It contains the original gems and cultured pearls. The cultured pearls are baroque shape and of medium quality. Diamonds are full-cut, indicating, perhaps, a better overall quality ring than assumed at first glance.

MODEL APPRAISAL NARRATIVE

Writing a Description of a Strand of Pearls

4–96. Strand of cultured pearls.

One (1) strand of round cultured pearls approximately 30 inches long, including the clasp, consisting of one hundred and four pearls (104), measured by pearl measuring gauge as 6.0–6.5mm in diameter. The strand is of uniform size beads.

The pearls have a white body color with slight rosé overtone. They are slightly spotted and have a high degree of luster.

The pearls have an approximate 1mm nacre thickness. Fair orient.

The pearls are well-matched for size and roundness on a knotted strand. The strand terminates in a 14 karat, stamped and tested, white gold fishhook-style filigree clasp.

The overall weight of the strand is 37.25 grams.

Current Replacement Value $________________

____

Appraiser

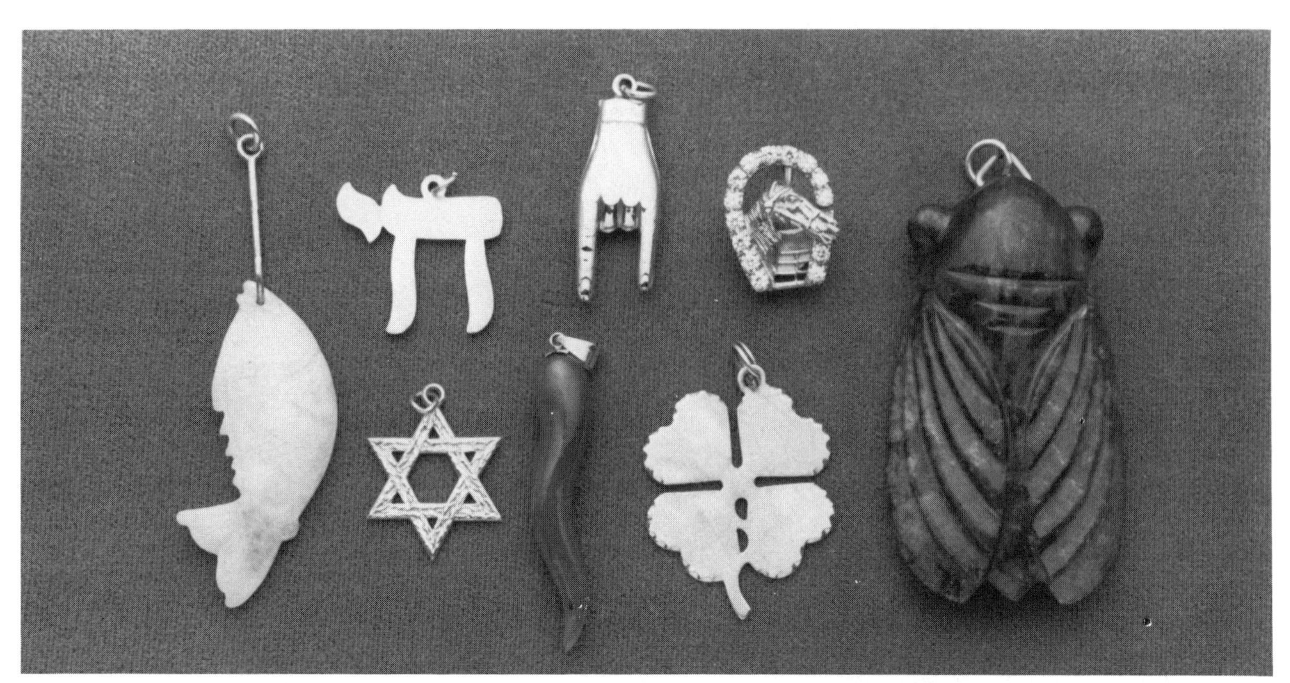

4–97. Charms worn to protect and to bring good luck are indigenous to all cultures. *Left to right:* The carp is an Oriental symbol and talisman for luck and endurance; the Chai and Star of David are religious charms; the hand, used as talisman, has been found in Etruscan tombs dating from 800 B.C. This form, with the second and third fingers closed and the first and fourth fingers extended, was considered an infallible protection against evil, particularly the Evil Eye. The horn is a talisman worn as protection from enemies and also to attract good fortune and success. Horseshoes were good luck symbols to ancient Greeks and Romans who nailed them to doorways with the horns upward as a charm against the plague. A four-leaved clover is universally believed to bring good luck. One leaf is for fame, one for wealth, one for a faithful lover, one to bring health! The cicada is an Oriental talisman for long life and good fortune.

4–98. Egyptian amulets and charms have esoteric meanings. The new pendant, *left,* is also the zodiac symbol of the archer Sagittarius. There is no older symbol than the scarab, which represents resurrection and eternity. The Eye of Osiris is worn to ensure health, courage, protection, strength, and wisdom. An amulet like the Anubis (a jackal-headed god, guardian of souls in the underworld) was worn to gain the favor of the deities. Ancient amulets in modern mountings like these three pictured may be costly.

Charms, Seals, and Trinkets

Charms

Charms were the first consideration of primitive humans, because they could be made of almost any small object—seeds, feathers, sea shells, bone, teeth, and claws. Because all the items were natural and made by unseen forces, they were believed to be powerful and have magical abilities to protect one from harm. Throughout history people have worn charms for the express purpose of bringing good luck.

What gradually evolved was a link between magic and religion, a complicated relationship between humans and supernatural powers. In many parts of the world today tradition is deep-rooted in the belief that common articles, including jewelry, work as charms. The terms *charm* and *amulet* (*amulet* is derived from Arabic and means that which is worn) are closely related, but the items are not quite the same. Although an amulet may be complete in itself, it is primarily a receptacle holding a charm symbolic in its influence to control good or evil. A charm supposedly exercises magic over one's health or even life (fig. 4-97 and 4-98). A good resource for charms such as these is Barakat Gallery, 429 North Rodeo Drive, Beverly Hills, CA 90210.

Charms are closely connected to pendants in the twentieth-century vernacular. During the 1980s we have seen a penchant for Egyptian charms and amulets in gold, including the Eye of Horus, which symbolized eternity, and the ankh, a symbol for long life. In the last few years, quartz crystal charms cut in pyramid shape and symbol-incised natural crystals have been popular as health and luck symbols. Although worn mostly for decorative value by the majority of purchasers in this country, this is not the case in many African, Middle Eastern, or Oriental parts of the world where the influence of gods and natural forces is accepted as fact.

The Christian cross has long been worn and revered as a protection against evil. It also symbolizes high civil and military honors. Stories rooted in old superstitions are rife with miracles said to have occurred thanks to the symbolic use of the cross. Splinters from the alleged true cross have been revered as relics for centuries and are preserved in some of the most impressive museums in the world. To touch the true cross was thought to bring reward. When you hear the term *touch wood* or *knock on wood* for luck, you will understand a direct connection to this ancient tradition.

In modern times there seems to be a lingering faith in emblems of good luck such as horseshoes, religious medals, chais, mizpahs, and figas. In the 1970s and 1980s, charms that have been cast in gold and worn grouped on bracelets or singly on chains or charm holders include four-leaf clovers, birthstones, wishbones, and lucky number charms. The use of charms as embellishments to a bracelet or neckchain may have started around 1910 when old seals bearing English coats of arms hung on bracelets.

Seals

The introduction of the adhesive envelope in 1844 was the beginning of the end for sealing wax and the functional use of seals. Until gummed envelopes were introduced, rings and seals with handles were common utensils for sealing personal documents and business correspondence. The early seals will be much larger than seals cut later, for the latter were meant to be worn as fobs with pocket watches. However, the earlier seals are finer artistic renderings and reflect the special abilities of the early engravers. As the utilitarian gave way to decorative use, seals were produced with engravings of a sentimental or romantic quality.

Classic intaglios that are made from gemstones are marvels of skill and patience and show intricate details of sculpture. These items may occasionally test the appraiser's research skills and patience. Antique jewelry dealer Myra Waller says that to determine metal content of seal mountings, examine inside the suspension ring atop the fob. This is a point of wear, and the base metal may be evident. This is also a suitable place to conduct acid testing.

It is becoming increasingly more difficult to find comparables, and the best source at the present time is a jewelry auction of antique jewelry in a major metropolitan area, such as New York or Los Angeles. The seals will need to be judged on their condition, rarity, age, material, material quality, quality of the carving, and the subject matter (simple or difficult). The seals will range in price individually from $250 to over $2,000 (fig. 4-99).

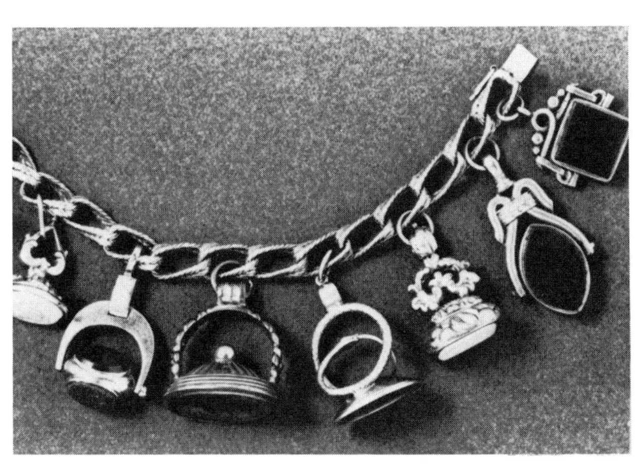

4-99. A bracelet with an assortment of seals and seal-fobs has been a popular jewelry ornament for years. One like that pictured can bring $3,000 at auction.

Trinkets

Trinkets are objects of luxury closely connected with costume, dress, and fashion. They have undergone transformation throughout the centuries from chatelaines to cigarette holders, as the social conditions shifted and cultural changes conformed to the times. The industrial revolution spoke both to the skill of the craftsman and to the fashion taste of the period. Women's emancipation contributed to the call for trinkets that reflected the new status of women, such as cigarette cases and holders. The old-fashioned chatelaines (metal clasps suspended by chains and furnished with hooks from which hung a variety of items such as scissors, pencil, and thimble) were common in the Middle Ages and revived in the early eighteenth century. It gave way to the hookless chatelaine worn with its two ends hanging from a tightly fitted belt, one end balanced by a seal and a tassel at the other. From this type of chatelaine, the single-purpose eyeglass case, money purse, or dangling watch evolved as accessory.

The trinkets of the twentieth century include cigarette cases by both unknown designers and famous and distinguished artisans like Fabergé. The burgundy-enamelled diamond and jade cigarette case in figure 4-100 by Fabergé is typical of his workshop. The piece is hallmarked with number *88* in an oval and *HW* (the workmaster Henrik Wigstrom). The Fabergé name in Cyrillic letters is in an elongated oval cartouche. The case has enamelling on guilloché background of silver mounted with chiselled gold leaf borders, 1½ inches wide, 4 inches long, and 1⅜ inches deep. It has a cabochon sapphire thumb piece. There are both old and new Fabergé reproductions, so the appraiser may need help from experts in the Fabergé field or consultation with a museum curator. Perusal of auction catalogs usually leads to sales information about comparable items. Two information resources are Forbes Galleries, New York, and the Walters Art Gallery in Baltimore.

4–100. A Fabergé cigarette box. Many reproductions of Fabergé jewelry and trinkets are currently found on the market. Distinguishing genuine from imitation is a test of the appraiser's skill. Genuine Fabergé items will have the name ***Fabergé*** spelled out in Cyrillic letters. All of the work will be precise and of excellent quality. Anything less was destroyed by the workmasters. (***Photograph courtesy of Designer Jewels***)

A trinket from the turn of the century that is regularly seen in estate appraisals is a miniature purse. These articles were approximately 2 inches in diameter, sterling silver, mesh, with cabochon sapphire thumbpieces. The little purse was used to hold coins in case a lady needed money during an evening out. It was generally suspended on a chain from an attached finger ring. They are valued from $200 to $300.

Questions and Answers About Charms, Seals, and Trinkets

Q. A collection of antique seals are to be donated to an appropriate museum. How can a receptive facility be located?

A. The search begins with perusal of the *Official Museum Directory* available in major libraries. The directory lists museums, their personnel, collections, research facilities, tax-exempt status, and so on in every state. This information helps donor find recipient.

Q. What is the best advice to give a collector of seals?

A. A methodical collection is always more valuable and noteworthy than a mixture of items. The serious collector should narrow his focus and specialize. A collection with continuity and with items of possible historical significance will be of greater value. For example, a wide assortment of Fabergé trinkets, from cigarette boxes to picture frames, is available. A notable collection would contain a number of cigarette boxes, authenticated, from either a variety of workmasters or from the hands of one particular workmaster. Specialization is a key to building an important collection.

Q. What are faux jewels?

A. *Faux* is a French term for false. Most costume jewelry set with clear or colored glass imitation gemstones is fashionably called faux. Prices can range from $35 to $400. Present popular styles are Art Deco to Retro with outrageously sized imitation gems, enormously sized drop earrings, and heavily ornamented brooches.

Q. What percentage of buyers at auctions are dealers?

A. Auction houses tell us that at Sotheby's and Christies dealers are 60 percent; at Skinner's of Boston, 75 percent dealers; at Butterfields in San Francisco, 50 percent.

MODEL APPRAISAL NARRATIVE

Writing a Description of a Charm

4–101. Four-leaf clover charm.

One (1) yellow gold charm stamped and tested as 10 karat, and set with one (1) full-cut round brilliant diamond, actual weight *0.10 carats.*

The charm is of die-struck manufacture in an open four-leaf clover motif assembled to a fluted stamped-out back. It measures 1¼″ h. x 1″ w. x ⅛″ d. (31mm x 24.5mm x 3mm). The diamond is bead-set in an illusion head in the center. Measurement 3.0mm SI-1. Color: J. Total gold weight, 4 dwts. The charm is stamped: HDC, 10K.

Current Replacement Value $______________

____

Appraiser

Cuff Links, Studs, and Stickpins

Clasps and buckles are old inventions, found in early times as strictly functional items of practical rather than ornamental use. They have turned into cuff links, shirt studs, and buttons in this century.

From biblical times to the eighteenth century men wore an abundance of jewelry. Arrival of the machine age and new levels of social status created an entirely new marketplace for men's jewelry, and etiquette of wear. Deprived of its symbolism, jewelry for men was confined to cravat pins, watches, watch chains with seals or fobs, and rings.

One hundred years ago, Griswold Lorillard introduced the tuxedo to men in Tuxedo Park, New York, when he appeared in a tailless dress coat, which shocked his peers. Lorillard had cut off the tails of his formal dress coat to lampoon the elder Tuxedo Park Club members who were obsessed with the English-style jacket. Thus, a new style was born and with it jewelry to accessorize the costume. The style has remained virtually unchanged since 1886: basic black trousers and dinner jacket with shirt studs and cuff links (fig. 4-102).

Cuff links remained unchallenged as a popular accessory to men's wear for several decades before going into a decline. In the 1980s, the style is making a comeback. We should see men's cuff links rise to new heights of popularity in the late twentieth century.

4–102. The ensemble pictured is a collection of men's jewelry with tuxedo studs, lapel pin, cuff links, and a key chain. (*Photograph courtesy of Diamond Information Center*)

The turnaround is caused by American men's attitudes toward fashion, which has changed dramatically. With more ways than ever to advertise their status and achievements, men are paying more attention to their appearance than ever before. Figures provided by the Men's Fashion Association indicate that American males spend approximately $40 billion a year on wardrobe and jewelry. Furthermore, MFA believes more men feel that it is now socially acceptable to wear jewelry other than a watch or wedding band. The thousands of men who purchased and wore neckchains, with or without medallions, in the decade 1970–1980 validate that belief. The brisk business in all types of jewelry for men (fig. 4-103), from card cases to collar tips, is going to offer expanded appraisal opportunities.

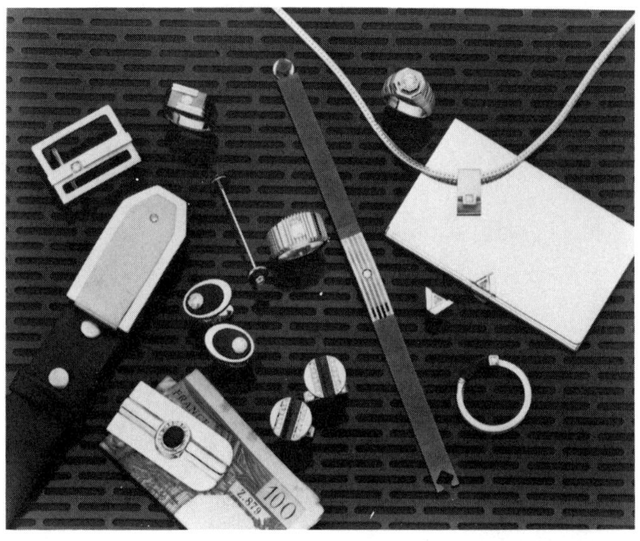

4–103. Men are wearing more jewelry in the 1980s, including precious metals on belt buckles and collar tips. The increase in men's jewelry wear will result in more appraisals. *(Photograph courtesy of Diamond Information Center)*

A Twentieth-Century Chronology of Men's Jewelry

1900–1910 Heavy gold pocket watches
Gold and gold-filled watch fobs
Gold stickpins with pearls
Cuff links and matching shirt studs with the most popular gem, pearls

1910–1930 Men's wristwatches appeared about 1914
Gold, gold-plated, and gold-filled collar pins
Tie clasps (appeared in 1914)
Collar buttons faded from fashion by 1921

1930–1940 Tie clasp with sporting motifs
Cuff links with colored gemstones
Chain tie clips
Key chains attached to a pocket watch. The chain was in a double snake-chain style with two strands twisted together in a cable twist.
Scarf pins with dogs or horse heads as ornamental subjects
Gold cigarette cases
Heavy gold link pocket watch chains

1940–1950 Large chunky chain identification bracelets for men
Large, chunky cuff links
1/2-inch-wide tie clasps
Palladium jewelry
Cigarette lighters in precious metals
Dog tags, gold- or silver-plated

1950–1960 Short tie bars
Wedding bands
Tie bar and cuff link sets
Key chains
Tie tacks
Colored stone and/or monogram rings

1960–1970 Tie clip out of fashion
Small link identification bracelets
Money clips
Rings, bracelets, pendants
Chains (with or without medallions)
Beads

1970–1980 Nugget jewelry (genuine and replica)
Cuff links/tie tacks out of fashion
Wedding rings, diamond rings
Colored gemstone rings
Variety of wristwatch styles and accessories

Stickpins

The stickpin was a popular item of jewelry in the early nineteenth century worn inserted vertically into a necktie, cravat, or cloth band folded about the neck. It kept the neck piece in place. The pins made in the eighteenth century had twisted or zigzag grooving to prevent slipping in the material and were shorter than those that followed. In the earlier models glass and foiled stones will frequently be found. The stickpins of

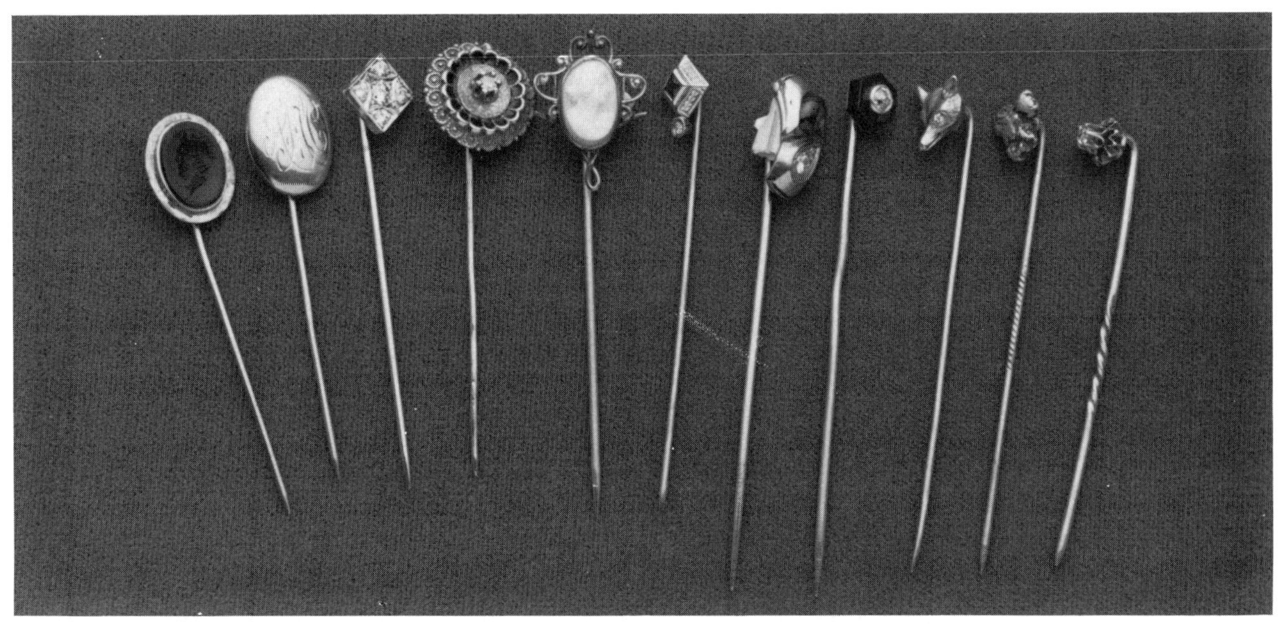

4–104. A variety of stickpins. Stickpins of the eighteenth-century had twisted or zigzag grooves and were shorter than those made later. Stickpins of the nineteenth century had grooving two-thirds of the way from the top. A man's stickpin would have had diamonds, colored stones, or a subject such as the fox mask or gold nugget. A woman's scarf pin would have had a cameo or monogram design.

the nineteenth century generally have less grooving with the metal twisting about two-thirds of the length near the top of the pin to secure the neck piece.

Some stickpins were made in pairs and worn together joined by a chain, with a safety device to prevent loss. Some had guards at the bottom to keep the pin point away from the wearer's throat. The stickpins from the early nineteenth century had three-inch-long pin stems with tops decorated and wrought from all kinds of metals, from base to precious (fig. 4-104), and in designs that depicted hunting scenes or other masculine activities. The stickpin has evolved into the modern tie pin or tie tack.

Women adopted the stickpin as an ornamental accessory about 1900 when the jabot pin, a type of pin similar to a tie pin and worn on a jabot (a ruffle on the front of a shirt), was popular. Women began to wear the stickpin on blouses or scarves, and the name *scarf pins* was coined. It should not be confused with the hatpin, a long pin meant to secure a lady's hat, used from Victorian times until the twentieth century. Both hatpins and stickpins have ornamental tops in various materials. In many styles, however, the majority of hatpins have base metal pins and are considerably longer than stickpins, often four to twelve inches long! They had nibs, small metal sheaths for the pointed end of the pin, that provided protection from the sharp end of the pin to both wearer and the public.

The major considerations in valuing stickpins are condition, materials, and age. They can have a value for insurance replacement from $50 for an unadorned example to thousands of dollars for one diamond-encrusted one. In this category, as in all others, if the stickpin has been poorly repaired (fig. 4-105) or had obvious lead solder repairs, it can lose up to 50 percent in value. Depending upon the extent of repairs, it may be worth no more than the intrinsic value of the metal or stones.

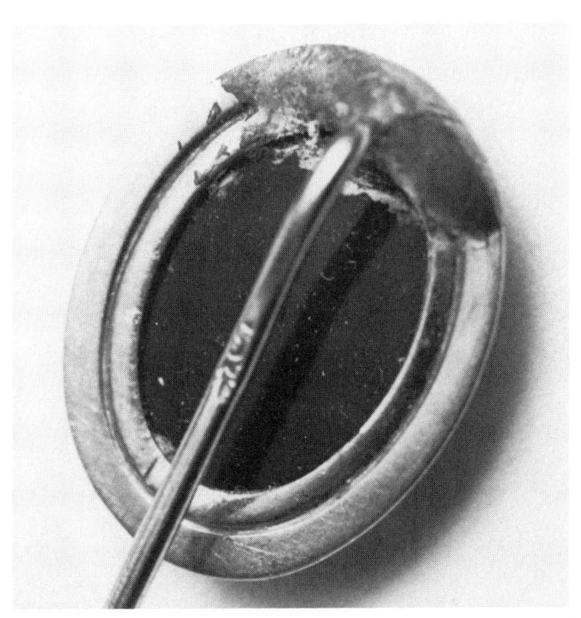

4–105. A poor repair job such as this will reduce the value of the stickpin 50 to 75 percent.

MODEL APPRAISAL NARRATIVE

Writing a Description of Cuff Links and Shirt Studs

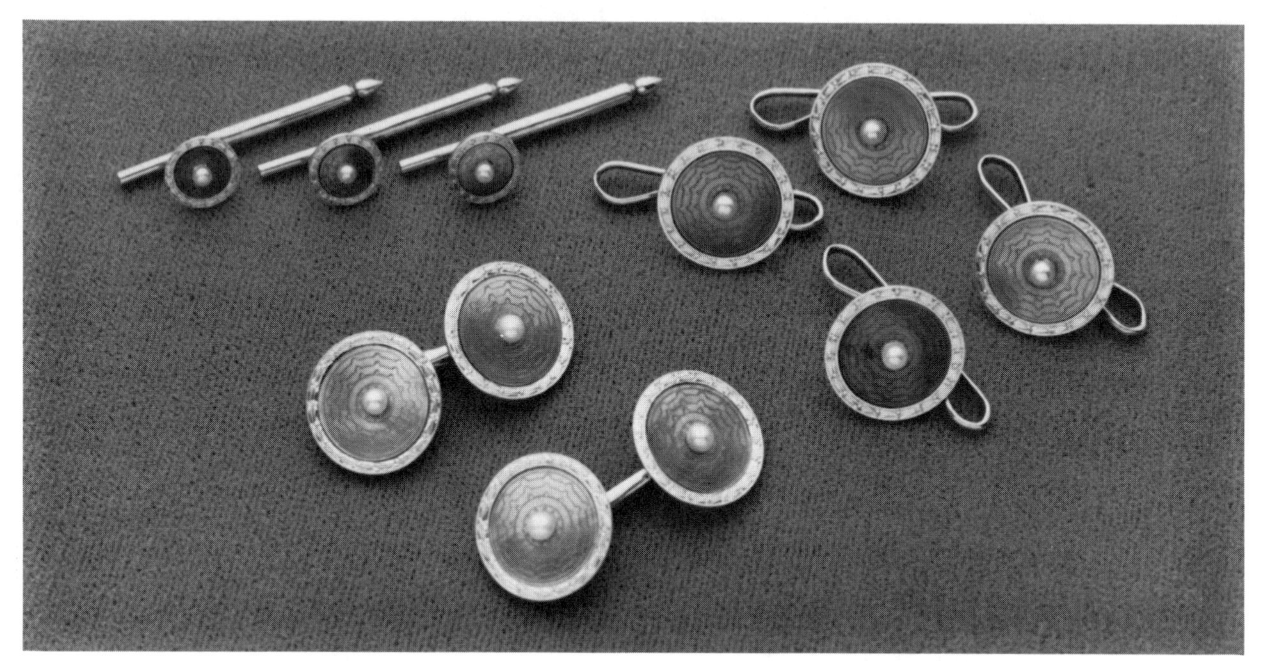

4–106. Matched set of cuff links, shirt studs, and vest buttons.

One (1) matched set of cuff links, shirt studs and vest buttons (9 pieces). The pieces are stamped and have been tested as platinum and 14 karat yellow gold combination.

The set, die-struck manufacture, consists of a pair of 14mm round double-faced cuff links, each side set with one (1) 3.2mm natural pearl. Four (4) 14mm round shirt buttons with toggle backs, and three (3) 7mm round vest buttons with spring backs; all are set with 3.2mm pearl.

The pieces are marked as solid platinum and 14 karat, circa 1910.

Total weight of the set, 24.50 grams.

All pearls set in the cuff links, shirt studs, and buttons are *natural pearls.* They were examined using X-ray techniques for positive identification in the XYZ gem testing laboratories of New York City, July 20, 1988. Certificate #P000/1606.

Manufacturer: The set is marked with the insignia of the manufacturing jeweler A. Grandé, founded in New York City about 1900 and still in existence. The company has been commissioned by several presidents of the United States to design jewelry articles for personal wear.

Provenance: This particular set of cuff links, shirt studs, and buttons were designed and especially prepared for a star of the theater and opera, Enrico Tenor, and was worn by him when he performed in *Rigoletto* in the New York Opera, October 1910. There are numerous photographs of Enrico Tenor wearing this set of jewelry. The jewelry is a gift from Mr. Tenor, by inheritance, to the present owner, his grandson. The owner (client of this appraisal) also has possession of the original order for the cuff links, shirt studs, and buttons, and the sales receipt. (Copies attached to this report for archival documentation.)

Current Replacement Value $________________

____

Appraiser

Questions and Answers About Cuff Links, Studs, and Stickpins

Q. What did the cuff links of the 1940–1950 period look like?

A. According to *Esquire* magazine, angled cuff links made an appearance in 1948. Cuff links were popular in plain or slightly domed styles. Colored stones, large sized and imitation gems, were popular set in cuff links. Another popular cuff link design was the knot, which today remains among the favorite motifs.

Q. Did designers sign stickpins in the 1920s?

A. Yes, occasionally. Those that are signed bring high prices, but the signature should not be overemphasized as an element of value. Numerous unsigned pieces have outstanding design and workmanship. Some fine stickpins of the 1920s were made of platinum, diamonds, and colored gemstones and were executed in either geometric patterns or a whimsical motif.

Q. When did the tie clip decline as a trend?

A. The tie clip was a fashion casualty of the 1970s, and its demise was the direct result of the wider necktie. The wide necktie was thrown off center when the tie clip was used. However, as more neckties were made with a loop designed to hold the short end of the tie in place, men became disinclined to punch a hole in an expensive silk tie, and the tie tack lost ground.

Q. What is Monel metal?

A. This was a trade name used for a nickel-copper alloy in the 1930s made by the International Copper Company. Monel metal had a silvery color finish and was often used in cuff links. The metal is seen with designs stamped into the surface.

Q. Precious metals are weighed in pennyweights; where did the *dwt.* abbreviation come from?

A. A convoluted scenario is the answer. The Latin coin denarius was a specific weight, and the Scots believed their penny weighed the same (as the denarius coin). The Scots adopted the troy system and took the denarius abbreviation *d* for the pence (penny) abbreviation. So, *dwt.* is a combination of denarius + weight = pennyweight.

Q. How is white gold alloyed, and when was it first used?

A. White gold was developed in the 1880s and was first used in the United States in the mid 1920s as a substitute for platinum. It is an alloy of gold, copper, nickel, and zinc.

Reverse Crystal Intaglio

Given half a chance, this fad of the 1860s could stage a big comeback in the 1990s. Reverse crystal intaglios are deeply carved cabochons of quartz crystal that have been painted and backed, usually with mother-of-pearl, so that the subject inside the crystal seems to be floating in space (figs. 4-107 and 4-108). What details! What distinguishes the antique crystal intaglios from early imitators and twentieth-century mimics are the incredible carving and painted realistic detail of eyes, hair, expression, blossoms, or petals inherent in the originals. This detail makes the pieces greatly appealing to collectors.

The crystal intaglio may be domed or shallow with a flat back, but the deeper the carving the more pronounced the trompe l'oeil effect. Motifs cover a variety of animal and floral subjects such as the fox, pet and hunting dogs (figs. 4-108 and 4-109), game and song birds, racing, riding and carriage horses, and every variety of flower (fig. 4-111). The domestic cat (fig. 4-110) and tigers (4-112) are in demand but difficult to find. Perusal of auction catalogs reveals that when the feline reverse crystal intaglio is offered for sale, it tends to bring much more than the price of other subjects, sometimes twice as much. Few sell for less than a thousand dollars. If you are asked to appraise this type of jewelry, research major auction houses for current indications of fair market value.

Dogs are popular subjects, especially Dobermans (invariably shown wearing a bright red collar), Jack Russell terriers, English pugs, terriers, and spaniels.

4–107. English reverse crystal intaglio brooch. (***Reginald C. Miller Collection***)

4–108. Dogs were one of the most popular subjects for the reverse crystals. (*Reginald C. Miller Collection*)

4–109. This type of intaglio was used in men's jewelry in the mid-nineteenth century. (*Reginald C. Miller Collection*)

4–110. Felines are one of the most popular subjects in the intaglio jewel. They are difficult subject items to find and when sold at auction tend to bring higher prices than other subjects. (*Reginald C. Miller Collection*)

4–111. The reverse crystal intaglios usually cost more than a thousand dollars. (*Reginald C. Miller Collection*)

4–112. Tigers are rarely found as subjects in reverse crystal intaglios. Examples like this brought over $3,000 at auction in 1987.

MODEL APPRAISAL NARRATIVE

Writing a Description of a Reverse Crystal Intaglio

4–113. French reverse crystal intaglio with bee, circa 1860.

4–114. Reverse of the intaglio brooch in figure 4–113.

One (1) yellow gold brooch mounting, marked and tested as 18 karat, set with a reverse crystal intaglio of a yellow and black bumblebee.

The handcrafted brooch mounting has bark-finish front, radiating-design back. The crystal has been drilled and secured to the mounting by four pegs. The brooch measures 1½ inches in diameter, 37mm. It has an extra long 18 karat yellow gold pin stem with a "C" clasp.

The crystal is round, high domed, and carved with the image of a yellow and black striped bee. Exacting details accent legs, wings, and antennae. The crystal is backed with mother-of-pearl.

Total gold weight with crystal, 15 dwts.

The pin stem is marked with an owl stamp, the French export mark. The piece can be circa dated 1860.

Current Replacement Value $________________

____

Appraiser

Reverse crystal intaglios, mainly found in men's jewelry, originated in 1860 in Belgium by Emile Marius Pradier. Thomas Cook developed it in England. The craft has long been identified with the famous jewelry firm of Hancocks in London. According to Harold Newman in *An Illustrated Dictionary of Jewelry,* the crystals are referred to as Essex crystal or Wessex crystal. This misnomer stems from the public's assumption that the crystals had been decorated by the famous nineteenth-century portrait painter William Essex.

The art was first used in stickpins but quickly spread to buttons, studs, and cuff links. After 1870 it was seen in women's jewelry, with colorful birds, butterflies, and insects as the most popular motifs. Insects were used for ornament at this time, especially the bee. This particular insect was greatly favored because it was the emblem of the Bonapartes.

An intaglio art mistaken for reverse crystal intaglio is the white sulphide relief in faceted glass. The intaglio sulphide technique was patented in England by Apsley Pellatt in 1831. This type of crystal, however, has no color and lacks details. The prices of the sulphides range in the hundreds of dollars. The colorful reverse crystal intaglio is found as a modern item listed in the Asprey catalog, a Fifth Avenue store in New York City; a pair of cuff links is priced at $1,700. One of the finest, most extensive collections of reverse crystal intaglios in the world is owned by Reginald C. Miller, gemstone importer and cutter in New York City. Miller's collection is museum quality, the result of several decades of search. He noted that the market value on the crystals seldom drops below a thousand dollars. Some can be worth several thousands of dollars depending upon the mounting. A few mountings are stamped with export hallmarks, but most are unmarked. Some mountings are inscribed on the back or have a container for a lock of hair. This attribute, the appraiser will remember, marks the piece as Victorian. A few crystals are designed as back-to-back pieces mounted to form a sphere in which the front and back of the subject is depicted. Fish, for example, seem to float in space and appear as three dimensional.

With a growing demand for artistic hardstone carvings, and the public's desire for fine cameos expected to increase, it seems reasonable that this lovely art form will grow in popularity as new collectors discover the craft. The appraiser must learn to recognize the original from the reproduction and the different levels of quality. Both Sotheby's and Christies have sold reverse crystal intaglios in the past few years. Some new reproductions were for sale at the annual Tucson Gems and Mineral show in recent years. Of concern to appraisers are the machine-carved items from Germany and Japan. Distinguishing machine mass-produced from handwrought, artistically designed requires field investigation and hands-on examination.

Pocket Watches and Wristwatches

When humans first began to measure time and carry the instrument of measurement about on the body is unknown. The ancient Egyptians correlated annual floods on the Nile with the stars and invented the calendar. The early Babylonians originated the concept of the sundial. Not until modern times, however, have humans developed an accurate timepiece (fig. 4-115). Some documentary evidence states that watches were first made in Nuremberg, Germany, by a locksmith named Peter Heinlein. He was the inventor of the coiled mainspring, about 1500. Others claim watch origination in Italy, Flanders, or Burgundy, but the questions as to where and when the first watch was actually made will probably never be satisfactorily answered.

One of the earliest existing watches, dated 1548 with the initials *C.W.* for Casper Werner of Nuremberg, is in the Wuppertal Museum in West Germany. The watch has a single hand indicating the time with both Roman and Arabic numerals. Another early watch, dated 1551, is in the Louvre in Paris. By the year 1600 watchmaking had spread to most of western Europe. Enough watches survive to enable us to trace a pattern of development. An interesting aside is that a ring on the watch case of the early models suggests watches were originally designed to be hung around the neck of the owner. Watches, rare and expensive, were meant to be seen and admired, as evidenced by a 1563 painting in the Royal Collection at Windsor Castle in which Lord Darnley is portrayed wearing a circular watch on a cord around his neck.

Early watch dials were engraved or enamelled with a scene, and the hands were simple. Since glass was not fitted to cases until after 1630, dials were protected by

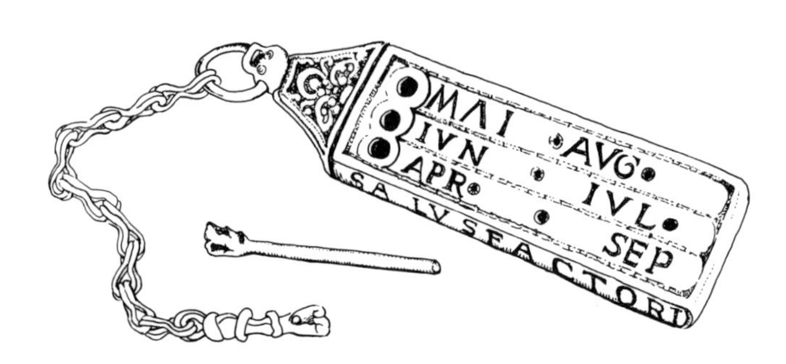

4–115. This is reputed to be the oldest watch in the English-speaking world. A Saxon, tenth-century pocket sundial, found at Canterbury in 1938 and now kept in Canterbury Cathedral. It is used as follows: Insert the pin in the hole appropriate to the month. Let the dial hang free facing the sun, so that the shadow falls down its column. At noon it will reach the lower spot, at 9 A.M. and 3 P.M. the higher spots. *(Illustration by Elizabeth Hutchinson)*

a lid or outer case. Case development followed an elaborate and decorative style set in France and Germany, with silver and gold gilt metal cases embellished with engraving and repoussé, and cases in fanciful shapes such as flowers, books, crosses, and skulls.

An impressive display of antique pocket watches covering many decades of development can be viewed in the Baltimore Art Museum. Although it is difficult to be able to appreciate the fine details of collector watches without being able to handle the pieces—in a museum collection this is forbidden—the value of such a grouping is the opportunity to see actual examples of rare period timepieces. This adds to the ability of the appraiser to identify quality collector items.

As a general rule, any watch dating before 1750 will be expensive and valuable. Watches that have lasted for over two hundred years must have been treated with respect by their owners. Much depends upon the watch's condition, decoration, movement, and dial. Watches from the 1775–1830 period have valuable movements but unimpressive cases. Watches from 1830–1900 may be quite inexpensive, as this was a period of many manufactured watches. The Colonial period of American watchmaking is pre-1850, but few American watches will be seen because so few were made. All American watches after 1850 are termed *modern period watches.*

According to experts and manufacturers, American pocket watches have not been made in the United States since the 1950s. The high period of production was the years 1850–1930. The most famous trademarks are names such as Howard, Waltham, Elgin, and Seth Thomas. The American railroad watch is among the most collectible of pocket watches.

The appraiser should keep in mind that, even though a small circle of dedicated watch collectors exists, antique pocket watches have cycles of interest for collectors that wax and wane just like those in antique jewelry. While precious metal weight influences the value, it is not as great an influence as in fine jewelry.

Condition is the number one factor in determining value of an old pocket watch. These are the main points to consider:

1. Condition
2. Brand
3. Date and genuineness of the period
4. Working order
5. Required repairs
6. Current demand

Research of realized prices for collector watches at auction show that antique timepieces have not fared well over the last ten years. Of course, there have been some extraordinary watches that brought special prices, but the message is that in an appraisal of most antique watches the appraiser has to determine more than just a date of manufacture and condition to estimate value. A thorough search in the current market for comparables plus consultation with dealers of old timepieces is paramount to a proper value statement. In any appraisal of pocket watches, the following points must be considered:

1. Decoration of the case usually increases the value.
2. Value increases if the watch is signed with the name of an eminent maker.
3. A bow badly worn or loose *may* lower value.
4. Dials with cracks, chips, patches, and so on, lower value.
5. Rusty hands may indicate damage below the dial and into the movement.
6. Plastic instead of glass crystal indicates replacement on an antique watch and lowers the value.
7. Worn cases with base metal visible lower value.
8. A watch in running condition is a plus factor that may raise the value. Gently rock the watch and see if it oscillates to-and-fro. If it has no wobble, the watch is in acceptable running condition.

Pocket watches are graded by their condition. The various conditions are:

- *Pristine,* a watch never used and in its original box
- *Mint,* one used but good as new with perfect dial, case, and movement
- *Average,* runs well and looks good but may have had part replacement
- *Fair,* a watch that is not in running condition but can be repaired (A watch deemed *fair* may also show a worn case or a small dent or scrape.)
- *Scrap* is a watch good only for its intrinsic metal content or its parts.

Fakes and Reproductions

There are a few problems with new fakes in old pocket watches, but not in great numbers. The reason for any fake is the opportunity for the perpetrator to make a lot of money. The fact is, it is just not profitable for anyone to invest time in faking antique pocket watches, unless they use the name of a famous craftsman. Most of the old silver or plated American pocket watches on the market today seldom appraise for more than $200 to $250. A gold-cased watch in good condition can be valued at $1,000 or over depending on the other elements in the watch. In the eminent craftsman category, it is assumed that the appraiser will be certain to establish that the age and serial number of a watch marked with a famous name are genuine so the valuer is not deceived.

Antique jewelry has married pieces, various components from different jewelry items put together to make one article, and so do watches. Pocket watches married to others with movements, dials, and cases are much more likely to be encountered than reproductions. Some points to look for that indicate a married watch:

1. Do the movement screws match to the watch case without signs of screwmarks in another place?
2. Are there spare holes in the dial plate indicating dial replacement? Do the screws align on the dial plate?
3. Is the dial a snug fit and not surrounded by an off-center gap around the case?
4. Does the three-part hinge that connects movement, case back, and glass bezel match?
5. Are there filled-in winding holes in the back of the case? If so, it may indicate an old key-wound case and replacement of the movement.
6. The wear marks on the cases should be even and constant, indicating two cases have been together a long time.
7. If the bow has been changed, there will be different sections of rubbings on the bezel.

A large quantity of information for identification and pricing is available from numerous books on the subject of pocket watches, catalogs, and illustrated literature. A source for watch and clock books is the Arlington Book Co., P.O. Box 327, Arlington, VA 22210. Another valuable resource is the National Association of Watch and Clock Collectors, Inc., 514 Poplar St., Columbia, PA 17512. The Association issues an impressive bulletin and valuable pricing information.

Wristwatches

Most estate appraisals today contain at least one wristwatch. It may be old or new, of base metal or precious metal, but unless the client is well versed in current pricing, or the watch is an enduring brand such as Rolex or Patek Philippe, he will have little idea of the current value. Fine designer brands, even older models, often bring high prices on the secondary market. Phillips Auction House estimates that period watches account for about ten percent of any watch sale. They point to the amazing record of Rolex, one of the most popular brands at auction, bringing twenty times today what the Rolex old steel models cost when new. The mechanical watches of the Art Deco period are in the middle of a revival as well as the Retro period watches (fig. 4-116). Enthusiastic bidding is reported by auction houses like Sotheby's for wristwatches with unique styling and fancy casework or unusually shaped watches. In early 1988, Sotheby's sold a gold calendar Patek Philippe Calatrava wristwatch, circa 1935, for $198,000. Reportedly it was a record for any wristwatch sold in the United States. The reason, according to the auction house, was that the particular case style had never before been seen in the United States. The watch had a simple calendar mechanism and fewer than five models were known to exist!

The first wristwatch on record was a present to Queen Elizabeth I of England in 1571. She wore it dangling on a chain from her wrist. This novelty item soon disappeared from the scene until the Empress Josephine commissioned a watch 250 years later. The fashion for wearing a wristwatch for men and women was firmly established during World War I. Today there is scarcely an arm in the world without a wristwatch.

Most watch appraisals, however, will not concern timepieces with prestigious heritage. Many will be mechanical and stem wind though, so it is important for the appraiser to note the following: most mechanical watches of value have a minimum of seventeen jewels. When you see twenty-one jewels in a wristwatch, it increases the accuracy of the timepiece. Some watches make a big show of having twenty-five or more jewels. In most cases, these extra jewels are purely decorative or auxiliary and are not working jewels. They add no extra value to the watch. Jewels are generally of synthetic corundum material, and their purpose is to act as bearings and reduce friction.

4–116. Deco, Retro, and late 1950s' model wristwatches are in demand, with many of the new owners replacing the mechanical movements with quartz movements for more convenient timekeeping.

WATCH WORKSHEET

Name____________________ Date__________

Type of case____________________

Case color Primary____________________

Secondary____________________

Case condition: Excellent Fine Good Worn Badly worn

Engraving Front____________________

Back____________________

Inside____________________

Comments:

Name on dial____________________

Dial type____________________

Dial color____________________

Dial condition: Excellent Fine Good Worn Badly worn

Comments:

Case manufacturer____________________

Case serial number____________________

Case quality____________________

Comments:

Movement manufacturer____________________

Movement size____________________

Movement serial number____________________

Movement manufacture date____________________approximately.

Movement type____________________

Jewels____________________

Movement condition:

Excellent Fine Good Poor Scrap

Comments:

4–117. Watch Worksheet. (*Courtesy of Edwin L. Menk, GG*)

A number of features of a wristwatch should be examined and noted on the appraisal worksheet and listed on the final report. Edwin L. Menk, horologist, has compiled the watch appraisal worksheet (fig. 4-117, see p. 119) to assist the appraiser. The following should also be noted:

1. Type of watch (man's or woman's)
2. Type of metal
3. Brand name
4. Condition (running, needs repairs, and so on)
5. Indicate if movement is mechanical or quartz
6. Number of jewels
7. Dial shape, color, type, condition
8. Embellishments on dial such as diamonds set as numerals
9. Style of hands and style of hour markers
10. Type of attachment, that is, strap or bracelet
11. Total metal weight
12. If gemstones are set on the case or attachment, cite the number, quality grade, and approximate weight
13. Embellishments to the dial such as moon phase, seconds dial, sweep second hand, and so on.

On the exterior, look where manufacturers may have sacrificed quality in case materials, or workmanship. Parts of the case may not fit snugly, which allows air, moisture, and dirt to enter the watch. Examine for plastic hands and dials.

Steven Orgel, of the New York wristwatch manufacturing firm Austern & Paul, says that he looks at the following when estimating quality: "See if the case is marked as waterproof, water resistant, dust proof, or humidity protected. Water resistance is the best element and to be marked as such must pass strict Federal Trade Commission rules." *Water resistant* is a term allowed by the FTC when a watch is immersed in water at normal pressure (15 pounds per square inch) for five minutes and then for five more minutes at a pressure of 35 pounds per square inch without showing any signs of admitting moisture. *Humidity proof* is not a term defined by the FTC. Orgel says it is loosely construed to mean a watch that is shower proof. *Dust proof* usually means that a thin metal band has been placed between the edges of the upper and lower plates of the watch movements to keep out dust. It is not an FTC term. Although *waterproof* implies that no water can possibly enter the watch, it is not an FTC term and therefore is open to interpretation. Orgel also instructs appraisers to look at wristwatch dials for signs of rust on the hands (indicating moisture and a possible broken seal), and to examine the watch back to see if it is a screw-in or snap-on back. "A low-quality watch has a back that can be flicked off with a fingernail," Orgel adds. "And a snap-on back that does not snap all the way around the case but only at two points—12 o'clock and 6 o'clock positions—is low quality." Other elements to watch for according to Orgel are the different types of movements and their manufacturers. A list of some of the best manufactured Swiss movements along with pertinent information about them can be obtained from the Swiss Jewelry and Watch Journal, P.O. Box 1345, CH-1001, Lausanne, Switzerland.

A trend today puts a quartz movement into a watchcase from the 1940s or 1950s to achieve the Retro look and eliminate the need to wind daily. This action may be validated by two points: parts for the mechanical movement may not be available; the old movement may not keep good time at any repair price. In estimating a value on this kind of timepiece, you have an altered item, and value is determined on that basis. Since the watch is a secondary market item, research the replacement price of the watch as sold on the secondary market (auction, second-hand jewelry store, antique jewelry) plus the price of a new quartz movement. Wristwatches in gold cases usually bring twice as much in the secondary market as the same watch in gold-filled, gold-plated, or stainless steel.

Currently popular are wristwatches decorated with plastic imitation lapis lazuli and malachite on dials and bands. Since the plastic looks natural to the unaided eye, magnification is needed for identification. Heft is a clue to the plastic imitator since the natural material is much heavier. The unpolished edges between links on expansion bands help identify natural material where the edges show a grainy texture, while plastic looks smooth. It is important to establish identity of the material, because the plastics are priced about $30 to $45 while the natural materials retail for over $100.

Wristwatch Reproductions and Knockoffs

The market is teeming with reproductions of famous brand-name wristwatches. Some are so obviously fake that you don't even need a loupe to tell the mimic, but some are less evident, in fact, sophisticated imitators. Rolex and Cartier are two brands with many imitators. In the Rolex brand wristwatch, many components can and are being changed. Get acquainted with the term *converted Rolex,* because it is becoming a familiar name to your clients. A converted Rolex can be changed in many ways. Generally it's a genuine Rolex movement with stainless steel case, replaced with 18K gold dial bezel and band and masquerading as a stainless-and-gold Rolex. To tell the mimic, look at the back of the case. A genuine stainless-and-gold model is marked at the end of a series of numbers with "18" or "14" signify ing the gold karat with the stainless. Other *conversions* imitate the Presidents model wristwatch.

4–118. The genuine Rolex is the watch with the winding stem pulled out slightly. Two Rolex imitations sit side by side. The center Rolex imitation looks good, but it is not as well made as the mimic next to it (to the right). One dial element to check is the Swiss label at the bottom. Notice that the genuine Rolex has a considerably smaller label, and it is farther down on the dial, almost under the bezel.

4–119. Two women's wristwatches. One is a genuine Rolex, one a counterfeit. The real Rolex brand is on the left. Close inspection will reveal many differences.

The cheapest knockoffs of all well-known brand watches can be found for sale on the streets of major U.S. cities and cost from $20 to $35 depending upon your bargaining skills. Better-quality imitations can cost up to $150. To spot the imitations, examine the dial. Except for the most costly counterfeits, the dial is poorly made (fig. 4-118, p. 121). Brand names may be smudged, crooked, or the print may be of a different style or size than that on the authentic brand watch. Logos are usually of a different size than on the genuine brand watch (fig. 4-119, p. 121). Dial legends like *Oysterquartz* are alien to the genuine item. Better-made watches have crystals of mineral sapphire while many imitation watches use plastic.

Since the idea is to get the most for the least, mimic watches are made in the quickest, cheapest way possible. Production and finish are inferior. Look for rough edges on bezels or bracelet bands and lack of screws to set the band. Pins that are removed when links are taken out are often just crudely finished wire. For watches with strap bands, look at the quality and markings on the buckles and attachments. Some of the women's Rolex imitations do not have a sweep second hand; numeral indicators and hands are different from the genuine article. Look at case backs. Imitations omit the serial numbers and the backs are most often of the snap-open kick-snap variety.

Some of the reissues and lookalikes of vintage watches, such as Hamilton and Longines, are popular collector items. Many reproductions have quartz movements where the originals were mechanical. Care must be taken in appraising the reissue watch and consultation with a watchmaker and collector is recommended.

4–120. The subsidiary second dial is the small dial at the bottom. This is an example of a single-sunk dial.

Wristwatches can be one of the most perplexing items to appraise. Canvass dealers for current factual and sales information, and shop your regional market to get acquainted with the genuine brands. Auction catalogs can help in pricing of vintage wristwatches. The National Association of Watch and Clock Collectors previously referred to can be of assistance in recommending experts for identification.

Table 4-4. American Movement Sizes Lancashire Gauge.

Size	Inches	Inches	Millimeters	Size	Inches	Inches	Millimeters
18/0	18/30	0.600	15.24	2	1 7/30	1.233	31.32
17/0	19/30	0.633	16.08	3	1 8/30	1.266	32.16
16/0	20/30	0.666	16.92	4	1 9/30	1.300	33.02
15/0	21/30	0.700	17.78	5	1 10/30	1.333	33.86
14/0	22/30	0.733	18.62	6	1 11/30	1.366	34.70
13/0	23/30	0.766	19.46	7	1 12/30	1.400	35.56
12/0	24/30	0.800	20.32	8	1 13/30	1.433	36.40
11/0	25/30	0.833	21.16	9	1 14/30	1.466	37.24
10/0	26/30	0.866	22.00	10	1 15/30	1.500	38.10
9/0	27/30	0.900	22.86	11	1 16/30	1.533	38.94
8/0	28/30	0.933	23.70	12	1 17/30	1.566	39.78
7/0	29/30	0.966	24.54	13	1 18/30	1.600	40.64
6/0	1	1.000	25.40	14	1 19/30	1.633	41.48
5/0	1 1/30	1.033	26.24	15	1 20/30	1.666	42.32
4/0	1 2/30	1.066	27.08	16	1 21/30	1.700	43.18
3/0	1 3/30	1.100	27.94	17	1 22/30	1.733	44.02
2/0	1 4/30	1.133	28.78	18	1 23/30	1.766	44.86
0	1 5/30	1.166	29.62	19	1 24/30	1.800	45.72
1	1 6/30	1.200	30.48	20	1 25/30	1.833	46.56

MODEL APPRAISAL NARRATIVE

Writing a Description of a Wristwatch

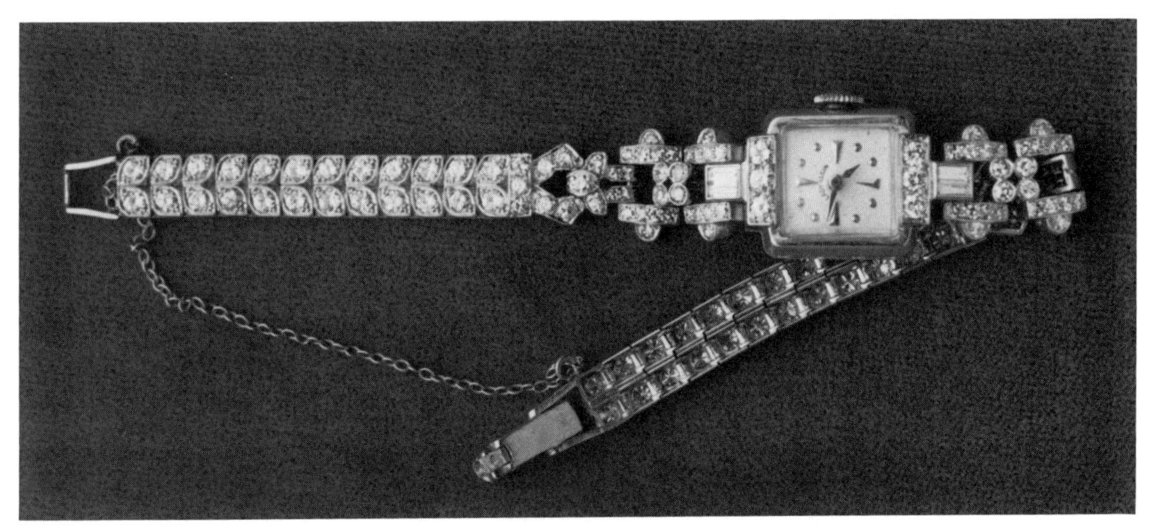

4–121. Platinum and diamond Elgin wristwatch.

One (1) lady's platinum and diamonds seventeen (17) jewel Elgin wristwatch with a platinum and diamonds bracelet band with fold-over clasp and safety chain, circa 1962.

The metal is platinum (stamped and tested). The square white dial is signed Lady Elgin and has stick-and-dot numeral markers and Dauphin-style hands. The mechanical movement is numbered as #330003.

The case is numbered as #2142. The serial number is #5112.

The watch case contains six (6) full-cut round brilliant diamonds, 3.0mm each, and four (4) square-cut diamond baguettes, 2.25mm each, having G/H average color, VVS-1 average clarity grades, and a total estimated weight by formula of *1 carat.* The bracelet contains 124 full-cut round brilliant diamonds, 2.37mm each, set in millgrain settings. The diamonds have an average G/H color grade, VVS-1 average clarity grade, and a total estimated weight by formula of *6 carats.*

Total weight of the watch 22 dwts. with movement.

Current Appraisal Value $__________________

____

Appraiser

Questions and Answers About Pocket Watches

Q. What does the term *subsidiary seconds* mean used on a pocket watch appraisal?

A. This is the small dial (fig. 4-120) on the face of the watch that has a single hand counting the seconds from 1 to 60. The watch may also have a sweep second hand.

Q. How can I tell if a pocket watch was for a man or a woman?

A. Most men did not wear a pocket watch that was under a size 12. Table 4-4 is a list of the American Movement Sizes, Lancashire gauge, that will help to find the size of a pocket watch and to use the correct size nomenclature.

Q. What do single sunk and double sunk mean?

A. Single-sunk dial means another dial (usually the seconds dial) at a different level. Double-sunk dial means a dial with separate portions at different levels, inside the hour-and-minute circle and for the subsidiary-seconds circle.

CHAPTER 5

THE APPRAISAL DOCUMENT

The culmination of examination, research, and analysis of jewelry items is the appraisal report. The thoroughness with which the appraiser has gone through these three phases gives weight to the value conclusions and instills confidence in the client. It also separates the tradesman from the professional.

The information contained in the report and the way it is packaged is your product. It should be merchandised deliberately with the use of special forms, document covers, preface pages, and impeccably typed narratives. Narrative descriptions form the heart of the work and explain relationships to various properties. The document is actually part of the history of the object. It is fitting for the appraiser to point out the obscure details of an object in the narration, spelling out what is not legible in the photographs. These details enhance the readability of the appraisal report, help to emphasize the importance of the report, and serve to protect the client in the event of loss of the article.

The American Society of Appraisers, the largest and oldest multidisciplinary appraisal organization in the United States, requires the inclusion of certain specific explanations, descriptions, and statements in appraisal reports issued by their members who offer their services for a fee to the general public. The required explanations, plus necessary additions include:

1. Letter of transmittal to the client
2. Document cover
3. Cover letter explaining purpose and function of the appraisal; most relevant and appropriate market; basis of the value conclusions; the state of the market at the time of the appraisal; and date of value (last date appraiser viewed the item)
4. Market definitions, that is, fair market value, retail replacement value, or other
5. Body of the appraisal report
6. Gemstone plots
7. Photographs
8. Glossaries or vocabularies
9. Contingent and limiting conditions to which the appraisal findings are subject
10. References or resource page
11. Statement of the appraiser's disinterest
12. Signatures on appraisal reports and inclusions of any dissenting opinions
13. Qualifications of the appraiser.

Information to Be Included in a Jewelry Appraisal

A basic inventory of information necessary to the preparation of the jewelry report includes:

1. Record of jewelry type, style, motif, and method of manufacture
2. Record of metal type, color, fineness
3. Weight of item in pennyweights or grams; specify whether gemstones are part of the overall weight
4. Record of jewelry measurements
5. Note of outstanding ornamentation and jewelry finish
6. Record of jewelry condition
7. Trademark, maker's mark, or hallmark
8. Description of the design and clasp(s); use of accurate nomenclature for settings
9. Identification and quality grading of gemstones with weights and measurements and designation of the grading system used.

Add the following information for an antique or period article of jewelry:

1. Circa date
2. Motif
3. Design
4. Assessment of repairs (good, average, poor)
5. Review of needed repairs
6. Provenance

An appraisal worksheet (fig. 5-1 below) is given to aid the appraiser in covering all points vital to estimating value in antique and period jewelry. The following report critique is used to assist in report writing, pointing out areas that must have fully narrated information and descriptions.

Appraisal Document Critique

The report shown on page 127 is based on a hypothetical appraisal document. The purpose of the critique (see p. 128) is to point out inaccuracies in form, to suggest information that should be included, and to cite specific omissions of material mandated by gemological and appraisal socieites.

SAMPLE ANTIQUE JEWELRY WORKSHEET

Jewelry Type__________

Method of Manufacture__________

Metal__________ Design__________

Motif__________ Style__________

Circa Date__________ Measurements__________

Embellishments (enamel, cannetille, etc.)__________

Workmanship (Excellent, Average, Poor)__________

Hallmarks or maker's marks__________

Clasp__________ Karatage__________ Tested__________

Repairs Made__________

Repairs Needed__________

Total metal weight__________

Gemstones__________

Identification/Gemstones__________

Quality grade(s)__________

Jewelry provenance__________

Jewelry Description__________

References: Sources__________

5–1. Antique and period jewelry worksheet. (*Courtesy of James V. Jolliff, MGA*)

JEWELRY APPRAISAL

Appraisal for: ______________________

Address ______________________

____

These estimated replacement costs are based only on estimates of the quality of the stones (unless specifically stated that the stones were removed and graded). We assume no liability with respect to any action that may be taken on the basis of the appraisal.

Date ______________ (1)

DESCRIPTION OF ARTICLE	Estimated Replacement Cost—Including Tax
(2) One lady's yellow gold (3) blue sapphire and diamond ring, (4) stamped 14K and weighing approximately 5 pennyweights. (5) This dinner ring is set with sixteen round, (6) faceted, (7) blue sapphires on the lower tier of the ring with a total weight of approximately 1.68 carats, nine round full-cut diamonds (8) with a total weight of approximately 0.71 carat and one oval-shape faceted (9) Ceylon blue sapphire in the center of the ring. The sixteen blue sapphires in the lower tier of the ring have an even body color of medium to dark blue. (10) The nine diamonds in the middle tier of the ring have an average clarity of VS-2 and an average color of "G" to "H" (est.). The large Ceylon blue sapphire which constitutes the upper tier, measures approximately 9.00mm length, 7.00mm width, 5.20mm depth, (11) and weighs approximately 3.08 carats. (12) It has an even body color of light to medium blue (13) and its clarity is LI-2 (est.) (14)	

Appraiser

Critique:

1. The report does not state a *Purpose* and *Function.*
2. The style is jumbled and difficult to read. Separation into several paragraphs would present an easier to read document. Each segment should be grouped in an individual paragraph.
3. Identify the large sapphire as natural or synthetic.
4. Test for karat fineness; don't depend upon a stamped mark.
5. Note if this includes gemstones.
6. Identify the small sapphires as natural or synthetic.
7. Note if sapphires are natural color or color enhanced.
8. Give the measurements, weights, and quality grades in GIA terms.
9. Country of origin should not be used to describe color.
10. Measurements and weights should be recorded.
11. Measurements should be written in the standard manner (9.00mm x 7.00mm x 5.20mm).
12. Note the way in which the weight was estimated. If estimated by GIA formula it should be noted on the document. Note quality grades.
13. An inconsistent sentence stating an *even* body color and describing an uneven color. Better to use saturation and hue.
14. State the color grading system used.
15. This appraisal has no description of the mounting, a vital part of the jewelry report.
16. The mounting should be described as to design, style, and motif.
17. The mounting *condition* as well as the overall condition of the ring has been neglected.
18. No mention of the type of manufacturing in the mounting.
19. No mention of hallmarks or trademarks.

One of the major problems in the settlement of jewelry insurance claims is the fact that many losses are not total and, therefore, the loss must be adjusted. A thorough, positive, and clear description of the mounting separate from the major stone(s) saves a lot of argument in the adjustment. Some appraisers will even give a separate value for the mounting. This is a matter of the individual preference of the appraiser and is not applicable to antique jewelry.

Using Limiting and Contingent Conditions

It is true that you cannot negate your responsibility as an appraiser by weighing down the appraisal document with limiting conditions. It is also true that we live in a very litigious society and must protect our work and our reputations. Therefore, every appraisal regardless of purpose or client should have a list of limiting and contingent conditions attached to the report. In fact, the American Society of Appraisers mandates it as part of professional practice.

It is *not suggested* that each one of the following limitations/conditions be printed on *every* appraisal report. Rather, the appraiser must make logical judgments with the option to survey the list and use the limitation(s) appropriate to the purpose and function of the appraisal, and the individual circumstance.

Limiting and Contingent Conditions List:

1. This report is made at the request of the party named for his/her use. It is not an indication or report of title or ownership.
2. Unless expressly stated, the condition of the item(s) is good for its type, with deficiencies and repairs noted and recorded. Ordinary wear and tear common to this type of item is not recorded.
3. The value expressed herein is based upon the appraiser's best judgment and opinion and is not a representation or warranty, implied or express, that the item(s) will realize that value if offered for sale in the open market or at auction. Value(s) expressed are based upon current researched information available on the date of this report and no opinion is expressed as to any future or past value, unless expressly stated.
4. Where an appraisal is based not only on the item(s) but also on data or documentation supplied therewith, this report shall so state by making reference thereto and, where appropriate, attaching copies.
5. By reason of this report, I am not required to give expert testimony or to appear in any court or hearing, nor engage in post-appraisal consultation with the client or third parties except under separate and special arrangement and at additional fee. If testimony or deposition is required because of any subpoena, the client shall be responsible for any additional time, fees, and charges regardless of issuing party.
6. The contents of this report shall not be used for any purpose by anyone but the client without the previous written consent of the appraiser. Copies of this report may be made only upon approval of appraiser. Disclosure of the contents of this appraisal report is governed by the bylaws and regulations of the American Society of Appraisers with which the appraiser is affiliated.
7. The values expressed are based on current market information on the date made and no opinion is expressed as to any long-range future value nor as to any past value. The analysis in this appraisal is based upon current market conditions without

considering market influences of fashion, economy, or politics.

8. Unless otherwise stated, the weights of all mounted gemstones are estimated by standard Gemological Institute of America formulas. All measurements are approximate. Although state-of-the-art gemological equipment is used, the actual weight of the gemstones when unmounted may be different than the estimated weight by formula.
9. The quality analysis of gemstones is limited by their mountings and setting styles. Also, lighting conditions affect the analysis of color grading. Actual quality, color, and clarity grade of a gemstone may differ, if analyzed in a different light source or out of the mounting.
10. The grading nomenclatures used for diamonds, colored gemstones, and pearls are those used in the systems developed by the Gemological Institute of America.
11. The appraiser reserves the right to recall all copies of this report to correct any errors or omissions.
12. This appraisal has been prepared in conformity with the principles of appraisal practice and code of ethics of the American Society of Appraisers, and according to the *Uniform Standards of Professional Appraisal Practice.*
13. The statements attached to this report represent all services rendered for this appraisal only.
14. Diamond(s) and colored gemstone(s) were graded only to the extent the mounting(s) would permit and were not removed from their settings for exact weight or quality grades.
15. The values in this report are based on the range of researched prices prevailing for comparable jewelry items in the most relevant and appropriate markets: (*Name the markets*).
16. Fair Market Value, as defined and used by the Internal Revenue Service in its publications, was used in this appraisal. Within the meaning of that definition, the Fair Market Value reported is the price that would be paid by the greatest number of willing buyers from willing sellers, neither being under any compulsion to buy or sell in the market in which these items are most commonly sold to the public—the auction market, and the secondary jewelry markets—the sale being to an ultimate consumer who is buying for a purpose other than for resale in the purchased form.
17. None of the items of jewelry constructed with precious metal, either singularly or in any combination of metal, were tested by the standard acid test due to the constraints placed upon the appraiser at time of examination. The metal will be visually identified and described for its type of construction and content.
18. Stated values are given item by item unless clearly stated as being per lot. The total of individual item values shall not be construed as an appraisal value for the whole lot but merely as the addition of single values. Where values are given by lot, the value per lot is for the whole and no opinion is given as to individual or proportionate values within the lot.
19. The appraiser has no present or contemplated future interest in the subject of this appraisal report that might tend to prevent a fair and unbiased appraisal.
20. The appraiser does not have a personal or business relationship with the parties involved that would lead a reasonable person to question the objectivity and validity of this report.
21. The appraiser's compensation is not contingent on any action or event resulting from the analysis, opinion, or conclusion of value of this report. Nor is the appraiser's fee based upon any fixed percentage of estimated value in a report.
22. The appraiser has made a personal, physical inspection of the articles specified in this report.
23. The appraiser has a potential future interest in the appraised articles.
24. This update supplements the original appraisal report of ______.
25. The articles originally appraised have not been reexamined for this updated report. The original analysis, opinions, and conclusions of the original report were relied upon; only the values have been changed in response to current market conditions.
26. The valuation in this report has been prepared for insurance claims and adjustments and represents the replacement cost of comparable items on the date of this report.
27. Unless otherwise stated, all colored stones are presumed to have been subjected to a stable and possibly undetectable color enhancement process.

Fair Market Value Document

Value estimations for *fair market value* require different market research and use separate value explanation pages than the insurance report.

Fair market value estimates are made for individual items in a market recognized as appropriate for that item. It is also a good idea to include a page in the report with several comparables explaining the source to substantiate the estimated values.

In the physical makeup of the FMV appraisal document, the example pages 141 to 144 are appropriate substitutions for the Cover Letter and Appraisal Preface in the following Model Appraisal.

TRANSMITTAL LETTER

APPRAISER'S LETTERHEAD

Date

Mr. Andrew Austen
3300 Bonnie Avenue
Ericsson, PA

Dear Mr. Austen:

Enclosed is our appraisal for insurance function of a lady's gold, diamond, emerald, and ruby ring. A copy of this report is being included for the convenience of your insurance agent.

The retail replacement valuation was conducted in our office on December 18, 1988, for the function of insurance security and replacement.

The total retail value of the item is $______ and reflects the consumer retail price of this item in the most common and appropriate market on the date of valuation.

The value stated represents replacement, new, based upon a research of sales of jewelry of identical or comparable quality and kind. The figure does not include sales tax or any other charges that might be payable, and you may wish to take this factor into consideration when calculating your insurance needs.

The complete appraisal report attached to this letter of transmittal has a total of 10 pages with photograph of the appraised jewelry.

Thank you for the opportunity to meet your appraisal needs.

Cordially,

MODEL APPRAISAL DOCUMENT

(Cover Letter)

APPRAISER'S LETTERHEAD

The following appraisal report has been undertaken to determine the retail replacement value (new) of the diamond, emerald, and ruby ring described herein. This appraisal is to function as a value estimate for insurance. The value is representative of this time and market.

The estimated value of the appraised ring is: $__________

Retail replacement (new) is defined as the value that is the mode of current sales in the most common and appropriate market for new, like kind, or comparable jewelry. The valuation of the appraised ring is the replacement value at the current date in the local marketplace.

A variety of sources were consulted to obtain the value. The specific marketplace addressed is the one in which the ring is most commonly sold to consumers by retailers in new condition: retail jewelry establishment.

The measurements of all gemstones and weights are estimates. They were formulated using accepted industry methods. All diamond and colored gemstone grading reflects the grading standards of the Gemological Institute of America. All jewelry evaluations and appraisals are subjective, and estimates of value may vary.

Quality analysis of gemstones is limited by various types of settings, mountings, and lighting conditions. The quality grades might differ if analyzed removed from the mountings, or under different light sources.

Metals are identified by acid testing. Designer and manufacturer is identified from trademarks, hallmarks, or maker's marks.

Considerations in the final estimate of value include: Quality grades of diamonds and colored gemstones, metal karatage and weight, designer/artist/manufacturer, trademark, type of manufacture, quality of craftsmanship, design, physical condition, repairs needed or performed, supply and demand, and economic conditions in the current jewelry market.

The value expressed is based on current market information and no opinion is expressed as to any long range or future value, nor as to any past value. The appraiser has no present or future interest in the purchase and/or resale of the property appraised and the fee for this appraisal is not contingent upon values submitted.

Page 1 of 10 Pages

APPRAISAL PREFACE

Definitions of Value of Retail Replacement

Replacement Cost New: This is the current and prevailing researched market price necessary to replace an item of jewelry with an identical article of the same utility and quality.

Replacement Cost with A Comparable Article: This is the current and prevailing researched market price necessary for either an exact replication of the item appraised, or replacement with a property comparable in quality and utility.

Approach to Value

Value estimates have been determined by using the market data comparison approach. The cost approach and revenue approach to value were not applicable.

Valuation Conclusions Based on Market Comparison

Following examination, identification, and qualitative assessment of the property described in this report, valuation conclusions were based upon comparison with:

Reported sales
Offers to sell by catalog
Retail jewelry store prices
Current manufacturers' price lists
Historical sales data in the appraiser's files/data bank

The utmost care has been taken to evaluate the maximum number of comparable properties in an appropriate market.

Jewelry Insurance

Homeowners' and Apartment Tenants' insurance policies commonly include coverage for personal property with the value of jewelry included up to a certain limit. Most policies restrict jewelry coverage for theft, burglary, or robbery to $500 aggregate, with a few companies providing coverage to $1,000 aggregate. Such insurance usually will not cover the property owner for "loss" of an item or for loss of a single stone from the item. In order to obtain coverage for the full value of the item(s), an appraisal for each item will be required and the coverage scheduled on your policy, with a specific description of the article and a replacement value on each item. For full and proper value coverage, a detailed professional appraisal is paramount.

In Case of Loss

Most insurance policies contain a clause which permits the company to replace the jewelry in like kind with a comparable replacement rather than make cash payment for the loss. If so, use your appraisal to verify that the quality of the replacement is similar to the lost item. The original appraiser of the insured item should be consulted for verification of the quality and value of the proposed replacement.

Appraisal Update

We recommend that your appraisal be updated periodically. Check annually with us about the necessity of updating your appraisal.

Standards

The analysis, opinion, and estimates of value conclusions in this report were developed and prepared in conformity with the Uniform Standard of Professional Appraisal Practice for the Personal Property Discipline of the American Society of Appraisers, and are in accordance with the Society's principles and ethics.

The appraiser is recertified as required by the mandatory recertification as set forth in the Constitution, Bylaws, and Administrative Rules of the American Society of Appraisers.

Certificate of Appraisal for Mr. Andrew Austen

ADDRESS 3300 Bonnie Ave.
Ericsson, PA

APPRAISAL TYPE Retail Replacement (New)
PRECIOUS METAL
BASE PRICE $450.00
DATE February 1, 1989

Purpose: Retail Replacement Function: Insurance

DESCRIPTION OF ARTICLE	ESTIMATE OF VALUE

Diamonds, Rubies, Emeralds, and Sapphires Ring—

One (1) ladies stamped and tested as 18K yellow gold ring containing twelve (12) diamonds, twelve (12) genuine rubies, ten (10) genuine sapphires, and four (4) genuine emeralds.

The 18K mounting is yellow gold and of cast manufacture. Heads for gemstones have been cast in place. The shank and top is of twisted wire design with a six (6) wire split shank. The top is hand assembled to the shank. The ring has an Arabesque motif.

The twelve (12) full-cut round brilliant diamonds in the ring are set eight (8) in a center rosette design, four (4) at the corners of the ring. Rubies, sapphires, and emeralds are set in flower (rosette) designs on four sides of the center diamond rosette.

The diamonds are:
One (1) full-cut round brilliant diamond measured by Leveridge gauge 6.50 x 3.90mm and weighing approximately *1 carat* estimated by standard Gemological Institute of American formula.
Clarity grade: VVS-1
Color grade: G
The diamond is set in a 7-knife-edge-prong setting.
Detailed plot analysis documented on page 7 of this report.

Page 4 of 10 Pages

DESCRIPTION OF ARTICLE	ESTIMATE OF VALUE
Eleven (11) full-cut round brilliant diamonds, 3.0mm x 1.9mm each by Leveridge gauge measurement. The weight of each diamond is approximately *0.10 carats* estimated by standard GIA formula, with the combined total diamond weight *1.10 carats.* Average clarity grade: VVS-1 Average color grade: G Each diamond is set in a 3-knife-edge-prong setting. Twelve (12) natural rubies. Each is round, faceted, and 3.0mm diameter. Depth was estimated using microscope and table gauge. Each ruby is estimated to be approximately *0.15 carats* with a combined and total ruby weight of approximately *1.80 carats.* One ruby is chipped The rubies are slightly purplish Red stones with medium to medium dark tone; light to moderately included with fair cutting, and approximately 40–60% brilliancy. Each ruby is set in a 3-knife-edge-prong setting. Ten (10) natural sapphires. Each is round, faceted, 3.0mm diameter. Due to setting restrictions the depth was estimated by using a microscope and table gauge. Each sapphire is estimated to be approximately *0.15 carats* with a combined and total sapphire weight of approximately *1.50 carats.* The sapphires are violetish Blue with medium to medium dark tone; free from inclusions under 10X magnification. The stones have fair cutting and approximately 30–60% brilliancy. Each sapphire is set in a 3-knife-edge-prong setting. Four (4) natural emeralds, round, cabochon-cut stones are the center stones in each "rosette". The emeralds measure 4.6mm diameter by Leveridge gauge each, and heights were estimated by using microscope and table gauge. The emeralds are approximately *0.35 carats* each with a total and combined approximate emerald weight *1.40 carats.* The emeralds are slightly bluish Green with medium to medium dark tone, light to moderate inclusions, and fair cutting. The stones have a 30–50% brilliancy. Each emerald is set in a 6-knife-edge-prong setting.	

Page 5 of 10 Pages

DESCRIPTION OF ARTICLE	ESTIMATE OF VALUE

The total and combined gem weight: *6.80 cts.*

Net gold weight of mounting, 10 dwts. The ring is size 7 with a shank that tapers from 4mm center back to 9mm at the shoulders

The ring is approximately: 1⅛″ h. x 15/16″ w. x 1¼″ d. (28mm x 23.5mm x 32.1mm)

The ring is trademarked 18K HB
HB is the trademark of Hammerman Bros, Inc.

Current Estimated Retail Replacement Value:

$________________

End of Appraisal

5–2. Diamond and colored gemstone ring (subject of appraisal report).

Appraiser

Instruments Used:
Binocular microscope
Leveridge gauge
Ultraviolet LW/SW
5-Diamond Master grading set
Pennyweight scale
Gold testing acids
Photographic equipment
Thermal conductivity probe-GIA
Duro-test Vita Lite
Refractometer
GIA or AGL Color Grading System

Page 6 of 10 Pages

DIAMOND APPRAISAL WORKSHEET

Name A. Austen | Job #
Address 3300 Bonnie Avenue | Date
City Ericsson, PA Zip | Date Due

Shape and Cut Full-cut Round Brilliant Loose Mounted x

Weight 1 Carat Estimated by formula Measured 6.50mm × 3.90mm

Diameter or Dimensions 6.50mm (length and width) Depth 3.90mm Table Measurement 4.30mm

Depth Percentage 60%

Table Diameter Percentage 66%

Girdle Thickness Thin to Medium

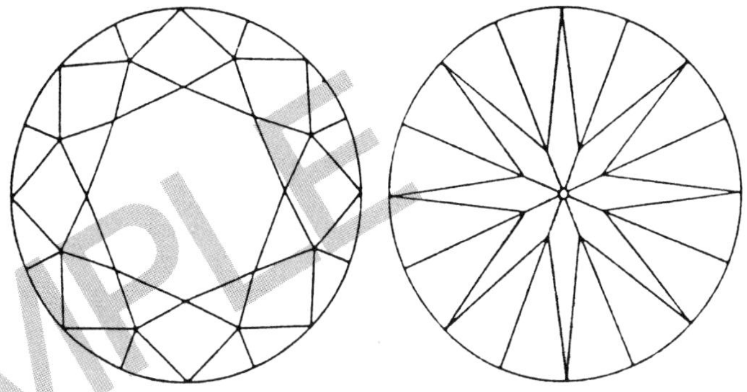

Finish
Girdle Surface Polished

Symmetry Good

Culet None

Polish Fair

Miscellaneous

Clarity Grade VVS-1

Color Grade G

Fluorescence None

COMMENTS:

Key to symbols:
. = pinpoint
` = feather

List Instruments Used:

Binocular microscope

5-Diamond Master Set (D-F-G-I-K)

Leveridge gauge - Ultraviolet

COLOR:

DIAMOND GRADING SCALE (GIA)

D E F G H I J K L M N O P Q to Z Fancy Yellow

COLORLESS NEAR COLORLESS TRACE COLOR VERY LIGHT YELLOW LIGHT YELLOW

CLARITY

Industrial Bort

FL VVS$_1$ VVS$_2$ Very Very Slightly Included — VS$_1$ VS$_2$ Very Slightly Included — SI$_1$ SI$_2$ Slightly Included — I$_1$ I$_2$ Imperfect I$_3$

Page 7 of 10 Pages

Color Grading Scale (GIA)

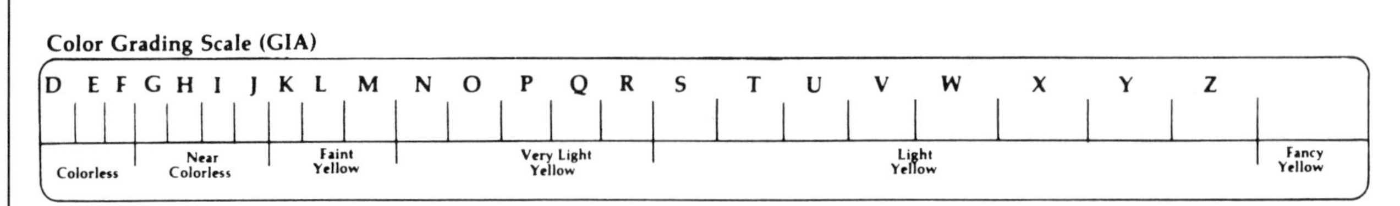

COLOR GRADE

The Diamond is assigned a color grade based upon the amount of body color it exhibits when compared to a set of Master Color Diamonds which have been color graded and registered as a set by the GIA Gem Trade Laboratory in Santa Monica, California. Under certain circumstances an Eickhorst Diamant Photometer may be utilized for color grading. All diamonds are graded using a standardized light source which is equivalent to North daylight.

Clarity Grading Scale (GIA)

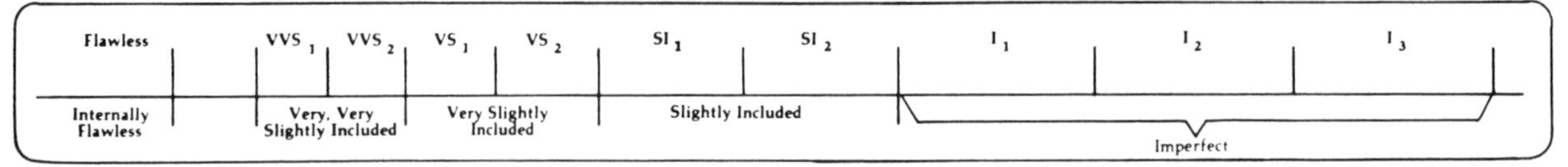

DEFINITIONS OF CLARITY GRADES

Consider the Size, Number, Position, Nature and Color of clarity characteristics when assigning the clarity grade of a diamond. Graders should sum up all these factors, rather than trying to analyze each separately.

FLAWLESS

Stones in this category are free from all blemishes or inclusions when examined in adequate light by a skilled observer with a 10x binocular microscope equipped with dark-field illumination.

The following conditions will not remove a stone from the flawless grade: 1) an extra facet located on the pavilion which cannot be seen from the face-up position; 2) a natural that does not extend beyond the maximum width of the girdle or break the symmetry of the girdle outline, and which cannot be seen from the face-up position; or 3) internal graining which is neither reflective, whitish, colored nor significantly affecting transparency.

INTERNALLY FLAWLESS

Stones in this category are free from all inclusions but do possess minor surface blemishes. Normally these stones may be made flawless by minor repolishing; exceptions to this general rule would be diamonds that exhibit surface grain lines, which could not be removed by repolishing. Such stones still might be graded IF.

VVS1 and VVS2 (Very Very Slightly Included)

These grades contain minute inclusions so small that they are difficult for a skilled grader to locate under 10x. Difficult to locate pinpoints, tiny feathers, minor bruises, or reflective internal graining characterize the VVS grades.

VS1 and VS2 (Very Slightly Included)

These grades contain minor or smallish in appearance inclusions clearly visible under 10x. Small included crystals, small feathers, and small clouds characterize the VS grades.

SI1 and SI2 (Slightly Included)

These grades contain inclusions which are noticeable under 10x. Stones in these grades (particularly the SI2 grade) may disclose inclusions to the unaided eye when placed table down on a white background. Included crystals, clouds and feather characterize the SI grades.

I1, I2 and I3 (Imperfect)

The imperfect categories contain inclusions which are quite obvious when viewed under 10x and which can be seen with the unaided eye. The grades vary from stones with inclusions so numerous that they affect transparency and brilliance to those with severe cleavages that are likely to be extended during ordinary wear.

TYPICAL CLARITY CHARACTERISTICS

EXTERNAL CHARACTERISTICS (Blemishes) PLOTTED IN GREEN AND BLACK

1. Natural — a portion of the original crystal "skin." Naturals may be located anywhere on a stone, but usually are on or near the girdle.
2. Extra facet — any polished surface in addition to the number of facets required for a particular cutting style and placed without regard for the stone's symmetry. Extra facets are often found around the girdle.
3. Pit — a very small opening on the surface of a facet.
4. Cavity, chip, or nick — terms used to describe an opening or hole on a polished surface. Nicks are usually found along the girdle edge.
5. Abrasion — minute chipping or white frosted appearance of facet junctions which results from one diamond rubbing against another or from blows during normal wear.
6. Scratch or wheel mark — surface marks caused either by another diamond being drawn across the surface of the stone or irregularities in the polishing lap.
7. Surface graining — visible surface indications of irregular crystal growth.

INTERNAL CHARACTERISTICS (Inclusions) PLOTTED IN RED

1. Feather — a term used to describe any break in a diamond (either cleavage or fracture).
2. Included crystal — a crystal of diamond or other mineral which is contained within a diamond.
3. Dark included crystal — an included crystal which is colored or appears black.
4. Knot — an included crystal of diamond which breaks the surface.
5. Pinpoint — a very small included crystal which ordinarily appears as a tiny white dot under 10x. Pinpoints may occur singly or in groups.
6. Cloud — describes any hazy of "milky" area seen in a diamond.
7. Internal graining lines — visible internal indications of irregular growth.
8. Laser drill hole — a man-made hole in the stone resulting from laser drilling, usually tubular in appearance. Laser drilling is normally done to lighten dark inclusions.
9. Bearded or feathered girdle — a girdle which exhibits hairline feathers extending into the stone.
10. Bruise or percussion mark — a crumbling or indentation in the surface often attended by rootlike feathers.

Proportion Analysis For Round Brilliant Cut Diamonds Only

Girdle Diameter (100%)
Table %
Crown Angle
Crown Height %
Total Depth %
Pavilion Angle
Girdle Thickness
Pavilion Depth %
Culet

RANGE OF ACCEPTABLE PROPORTIONS

Table Percentage	54 %	to	65 %
Depth Percentage	57.5%	to	62.5%
Crown Height	11.0%	to	16.0%
Pavilion Depth	41.5%	to	45.5%
Crown Angles	30.0°	to	35.0°
Pavilion Angles	39.7°	to	42.4°
Girdle Thickness	Thin, Medium, Slightly Thick and Thick		
Culet Size	Small, Medium and Slightly Large		

Page 8 of 10 Pages

Form courtesy of Tenhagen Gemstones.

LIMITING CONDITIONS

It is understood and agreed that fees paid for this appraisal do not include the services of the appraiser for any other matter whatsoever. In particular, fees paid to date do not include any of the appraiser's time or services in connection with any statement, testimony, or other matters before an insurance company, its agents, employees, or any court or other body in connection with the property herein described.

It is understood and agreed that if the appraiser is required to give court testimony or make any such statements to any third party concerning the described property or appraisal, applicant shall pay appraiser for all of such time and services so rendered at the appraiser's then current rates for such services with one-half of the estimated fee paid in advance to the appraiser before any testimony.

Financial responsibilities for this valuation report shall be limited to fees rendered.

The suitability and intended use of this appraisal in its entirety are predetermined; therefore, the format and values established are valid for the stated purpose of the appraisal only, and considered invalid if used for any purpose unknown to the appraiser. This report, or copy thereof, may be transmitted to a third party, or legal entity, only in its entirety.

The appraiser assumes no responsibility for unforeseen changes in market conditions. This appraisal is not an offer to buy.

MARKET DATA RESOURCES

Market References

The Guide, 1989
Rapaport Diamond Report
Fair Market Value Monitor, Annual Publication
The Diamond Yearbook, 1988–1989

Auction Comparables

Sotheby's

January 27, 1988	Lot 60	Price realized	$2,900.00
April 6, 1988	Lot 110	Price realized	$3,200.00
June 1, 1988	Lot 330	Price realized	$3,300.00

Butterfield & Butterfield

September 12, 1988	Lot 30	Price realized	$3,250.00

Christie's

September 22, 1988	Lot 46	Price realized	$2,700.00
October 1, 1988	Lot 60	Price realized	$3,000.00

Page 9 of 10 Pages

APPRAISER'S LETTERHEAD

Qualifications of the Appraiser

Education	Ph.D. Valuation Science (1983) Loretto Heights College, Denver, Colorado M.A. Valuation Science (1981) Lindenwood College, St. Charles, Missouri
Credentials	Graduate Gemologist, Gemological Institute of America Fellow, Gemological Association of Great Britain Master Gemologist Appraiser, American Society of Appraisers Senior Member, American Society of Appraisers
Business	Senior partner, APPRAISAL COMPANY, a professional corporation, New York, New York, with regional offices in Atlanta, Ga., Houston, Tx., Santa Fe, N.M., and Los Angeles, Ca. Seventeen years experience in the jewelry industry in sales, management, and appraisals.
Instructor	Lindenwood College Valuation Science Degree Program. University of Southern California Valuation Science Degree Program Rice University, Houston, Texas, Valuation Science Degree Program University of Houston, Valuation Science Degree Program Washington University, Washington, D.C., Valuation Science Degree Program
Biography	Who's Who in American Colleges and Universities, 1985 Who's Who in Business and Science, 1988
Expert Witness	United States Bankruptcy Court Federal Tax Court Family Law Courts Civil Courts

Page 10 of 10 Pages

FAIR MARKET VALUE DOCUMENT

(Cover Letter)

To: Kay M. Jones
#1 Matthews Avenue
Cody, Texas

Purpose & Background

This appraisal report was made at the request of Kay M. Jones for the dissolution of common property and is to be used for determining the fair market value of the jewelry articles listed herein. The monetary evaluation is based on the values expected in the market where such items are most commonly sold and the type, condition, and quality of the jewelry under consideration. This appraisal was prepared on February 1, 1989 and is effective as of that date. It is subject to the terms and limiting conditions listed in the report.

Procedure for Examination and Appraisal

This appraisal report was performed by a tested and certified gems and jewelry appraiser using the latest "state-of-the art" methods and precision equipment. All jewelry was examined and graded "on-site". Jewelry and/or gemstones described in this appraisal have been analyzed and graded using industry standards for diamonds, colored gemstones, and precious metals. Each item described has been researched for fair market value in its most appropriate market for that item.

As a general definition of fair market value, this appraiser considered the price at which the items would change hands between willing buyers and willing sellers, neither being under compulsion to buy or sell, and each having reasonable knowledge of the relevant facts. The specific markets addressed are the ones in which the items are most commonly sold to consumers by retailers who deal in secondary market jewelry.

Unless otherwise stated, the weights of all mounted gemstones were estimated by standard Gemological Institute of America formulas. All measurements are approximate.

The quality analysis of gemstones is limited by different mounting and setting styles, as well as lighting conditions. Therefore a range of quality may be given, that is, color of diamonds, G. The actual quality, color, and clarity of gemstones could be different, if analyzed out of the mounting or under a different light source.

Identification of metals and methods of construction are determined only to the extent that the item permits. When items are marked with the fineness of the metal, the appraised value is based upon the marking, unless otherwise stated. When there is no marking on an item, the method used to estimate the metal content is a field assay acid test. Only approximate results can be expected from this type of testing.

Page 1 of 4 Pages

Considerations made in the determination of dollar value include: weight, clarity, color, saturation, tone, quality of cut, gemstone desirability, designer/artist/manufacturer, trademark, hallmark, type of manufacture, quality of craftsmanship, design, physical condition, repairs needed or performed, supply and demand, and the present economic condition of the local market.

Grading nomenclature used for diamonds, colored gemstones, and pearls are those used in the system developed by the Gemological Institute of America. Diamonds are graded with the use of pre-graded permanent master diamond color comparison stones, and the grading nomenclature prescribed by the Gemological Institute of America. Colored gemstones are graded using the GIA prescribed colored stones grading nomenclature.

Determining Value

The fair market value of the described jewelry items has been researched with comparable sales of like items in the local and regional markets within the past 6 months period.

Fair market value is defined by the Internal Revenue Service (Revenue Procedure 66-49) and by Treasury Regulations as follows:

> Fair market value is the price at which the property would change hands between a willing buyer and a willing seller, if neither one is under any compulsion to buy or sell and if both have reasonable knowledge of all relevant facts. The fair market value may not be determined by the sale price in a market other than that in which the item is most commonly sold to the public. Thus, in the case of an item that is generally retailed, the fair market value is the price at which the item or a comparable item would be sold at retail.

For the purposes of this appraisal, the major markets meeting the complete qualifications under the fair market value definition are determined to be the secondary jewelry retailers, auction houses such as Sotheby's and Christies, and others operating in a way so as to have trackable comparable sales.

The values are based on current market information and no opinion is expressed as to any long-range future value nor as to any past value. Analysis is based on current market conditions without taking into account possible influences to the market caused by changes in fashion, economics, or political events which could occur after the date of this report.

Physical condition of the items is good unless otherwise stated in the detailed description.

Page 2 of 4 Pages

Use of Report

It is understood that the jewelry items in this report belong to Kay M. Jones and that the manner of acquisition can be formally stated.

This report is provided strictly for determining value for the specific purpose stated and at the specific market level stated herein. The procedures and wording of this report would be different for any other purpose, such as insurance replacement.

Date: ____________________

____
Appraiser

APPRAISAL PREFACE

Definition of Fair Market Value

The attached appraisal report has been requested for Fair Market Value (FMV), and a FMV has been estimated.

Fair Market Value is defined as "the price at which the property would change hands between a willing buyer and a willing seller, neither being under any compulsion to buy or sell and both having reasonable knowledge of the relevant facts. The value of a particular item of property is not the price that a forced sale of the property would produce. Nor is the Fair Market Value of an item of property the sale price in a market other than that in which such item is most commonly sold to the public, taking into account the location of the property wherever appropriate. Thus in the case of an item of property which is generally obtained by the public in the retail market, the Fair Market Value of such an item of property is the price at which the item or a comparable item would be sold at retail."
—U.S. Treasury Regulation 20.2031-1(b)

To evaluate an item for its cash value, a level of market must be recognized as appropriate for that article. The most appropriate market for jewelry can vary depending upon the article's age, condition, quality, intrinsic content, aesthetic appeal, and/or provenance. Other factors impacting market considerations are current fashion trends, artistic interpretation, period of manufacturing, and so forth.

Valuation Conclusions Based on Market Comparison

Fair Market Value is commonly estimated by using the market data comparison approach. The levels generally recognized as appropriate for this type of evaluation are:

Retail. Items in exceptional or new condition.

Wholesale. Fair market value at the wholesale level applies in those instances where the wholesaler is the *end user* of the product. For instance, large parcels of cut gemstones are generally utilized by a manufacturing jeweler rather than a retail jeweler or an average jewelry consumer.

Auction. This represents the average retail sales prices realized at major auction. The jewelry may be used but in good condition; of interest to collectors; has universal interest as an antique or period item; is an important item from a major retailer; was made by a recognized artist; has historical interest or provenance; is from a famous collection; is of exceptional quality or rarity.

Scrap. An item is unmarketable in its present condition and the value is based upon intrinsic content.

Page 4 of 4 Pages

APPENDIX

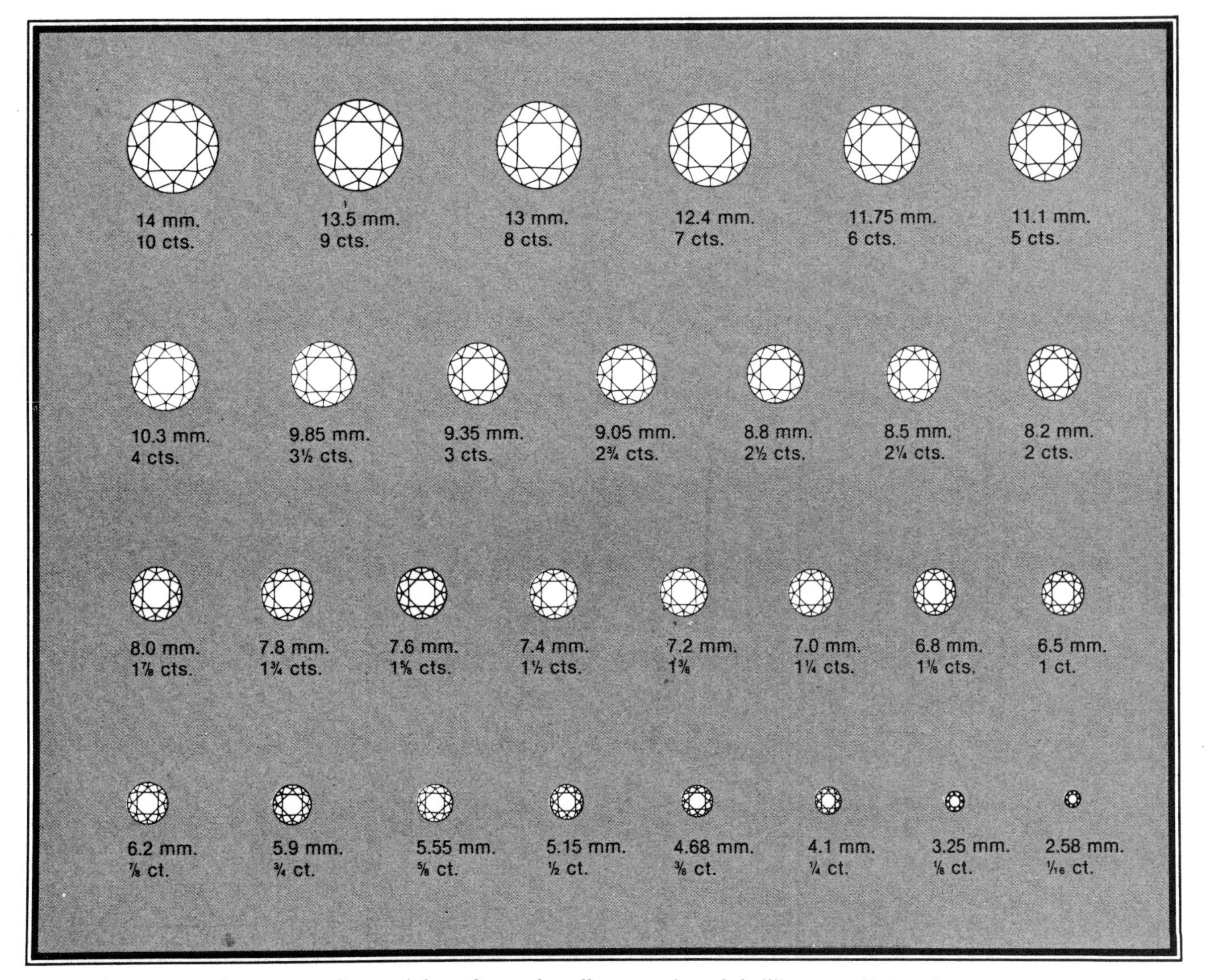

A–1. Diameters and corresponding weights of round, well-proportioned, brilliant-cut diamonds.

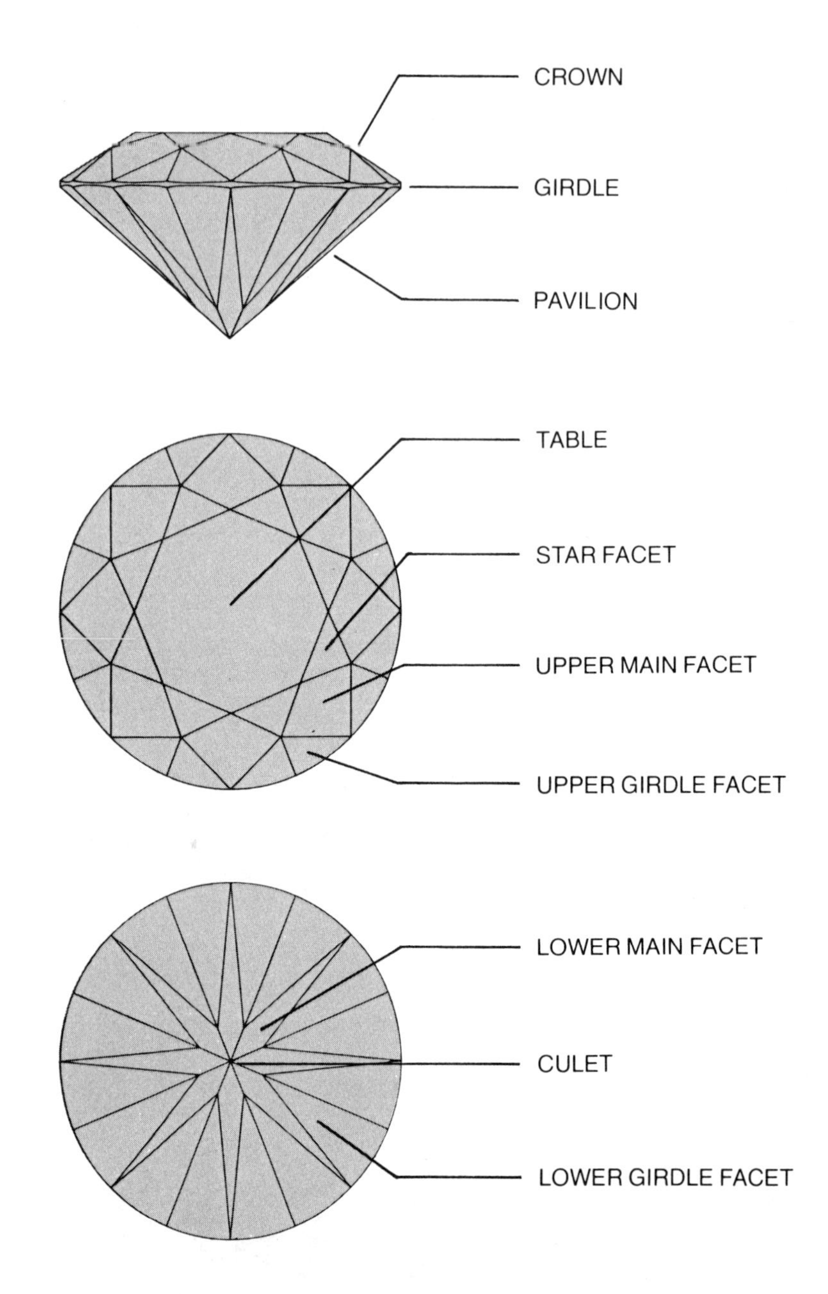

A–2. Proper nomenclature of the specific parts of a brilliant-cut diamond.

New Diamond Cuts

The Central Selling Organization (CS0) Chairman Mr. Nicholas Oppenheimer announced five new diamond cutting styles in 1988. The registered designs were created by Mr. Gabi Tolkowsky to maximize brilliance, color, yield, or a combination of all three. Heretofore, these rough stones were too difficult to manufacture profitably into the conventional round or fancy shapes.

The new cuts seen on the following pages are based on nonconventional angle dimensions. The overall proportions, as well as the greater number of facets around the culet, increase brilliance and improve the visual appeal of these stones. These cuts give a higher yield, an improvement in the face-up appeal, a greater consistency of color throughout the stone, and an increased brilliance (fire) over conventional shapes that are manufactured from difficult off-color roughs.

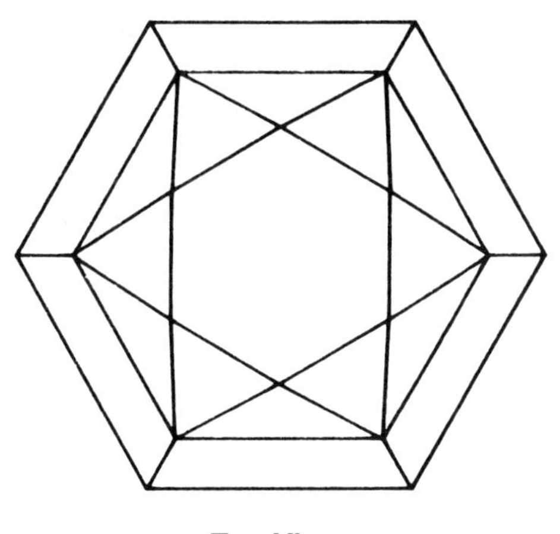

Top View

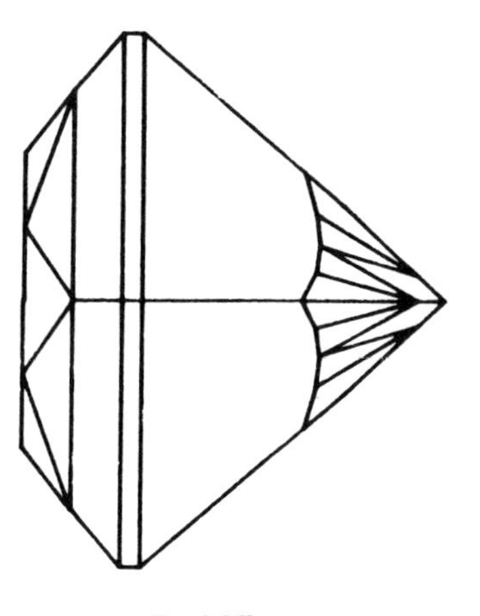

End View

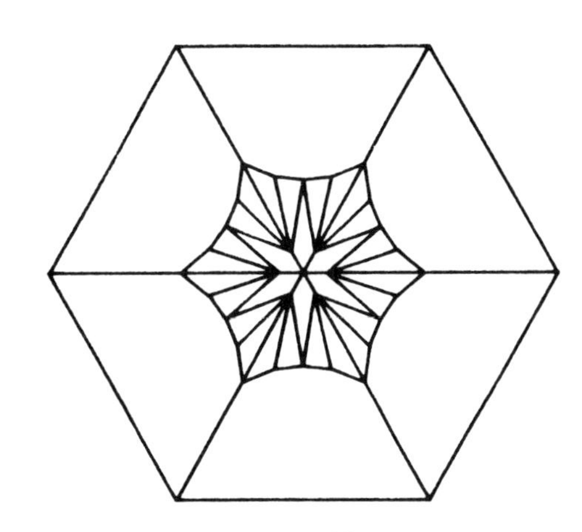

Bottom View

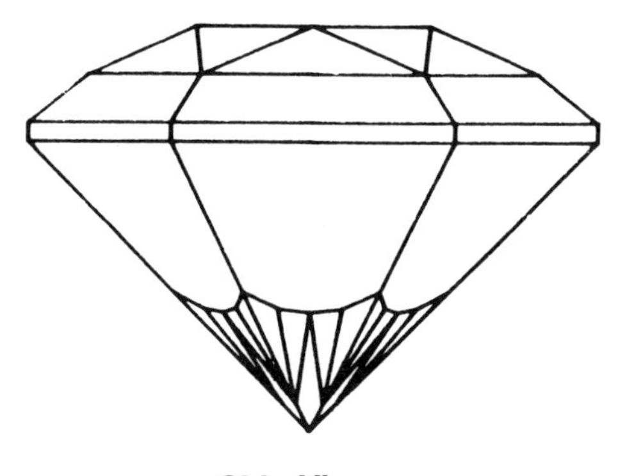

Side View

A–3. *Fire-Rose.* This cut is available as a round, pear-shape, marquise, and heart-shape. The advantages of this cut are a higher yield, increased brilliance (even in very strong colors), and an improvement in the face-up appearance of the stone in color as well as clarity. *(Information courtesy of Diamond Information Center)*

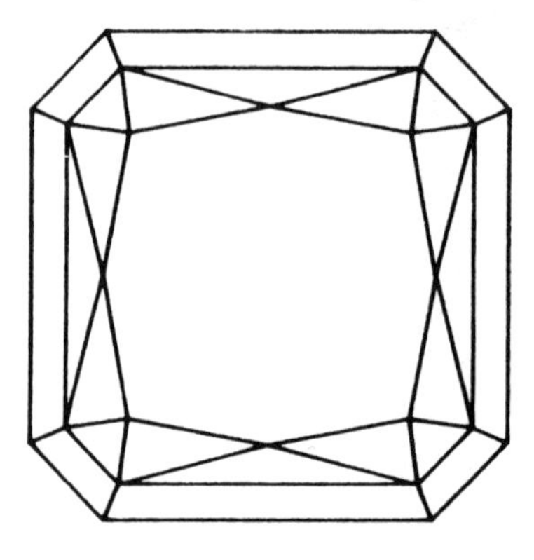

Top View

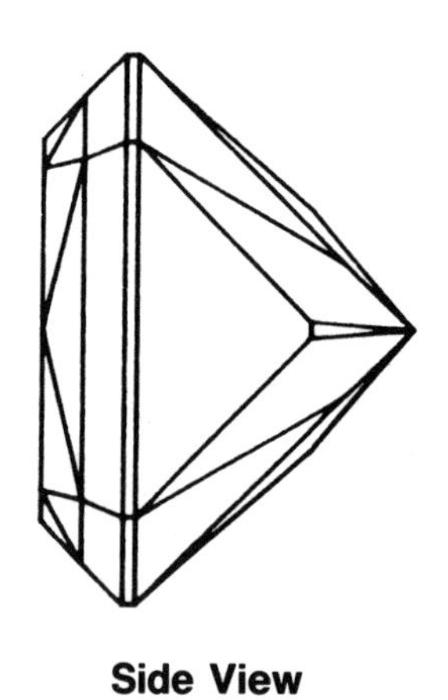

Side View

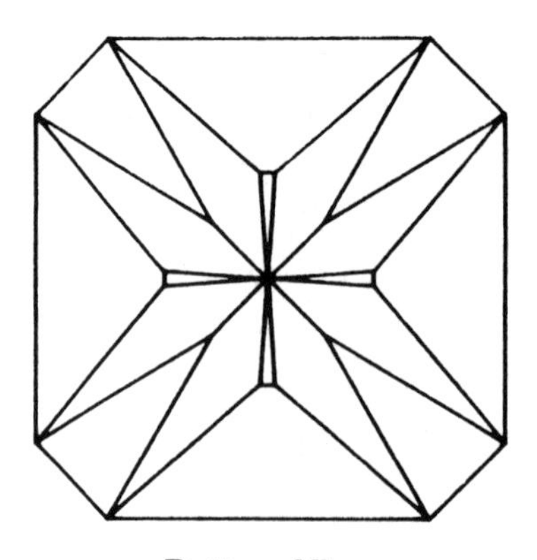

Bottom View

A–4. *Sunflower.* This is available as a carré, emerald, baguette, taper, marquise, pear-shape, and heart-shape. Advantages are the same as Fire-Rose. *(Information courtesy of Diamond Information Center)*

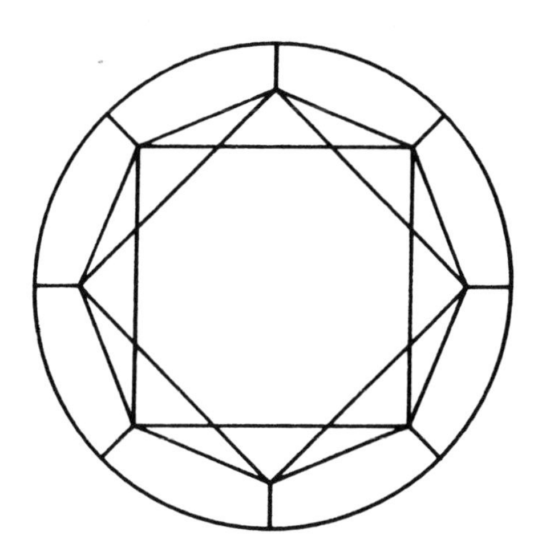

Top View

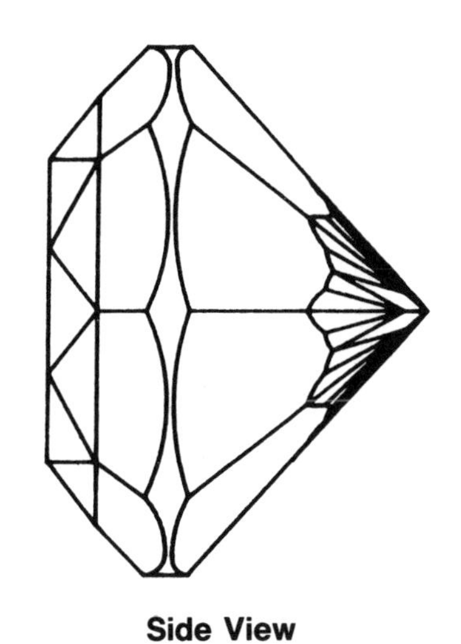

Side View

Bottom View

A–5. *Zinnia.* This comes as a single shape, which is a round stone with an increased number of facets around the culet, giving it a tremendous fire. Other advantages are the same as with Fire-Rose and Sunflower. *(Information courtesy of Diamond Information Center)*

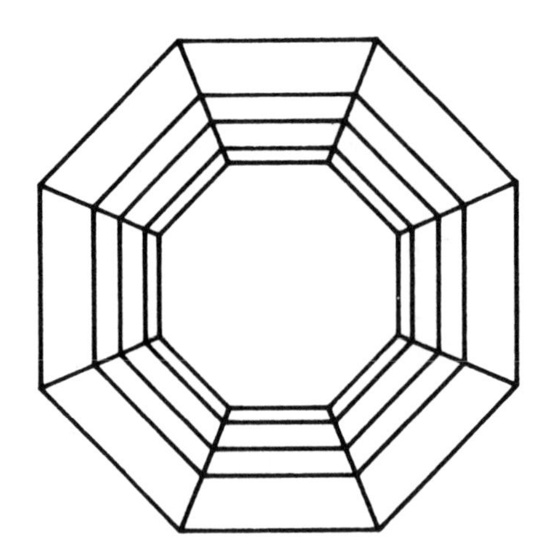

Top View

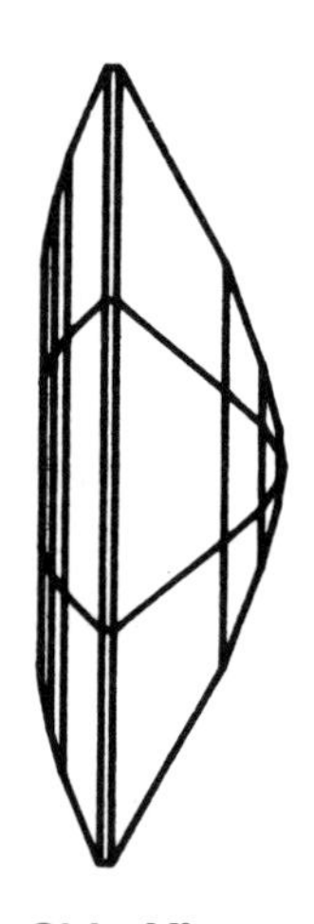

Side View

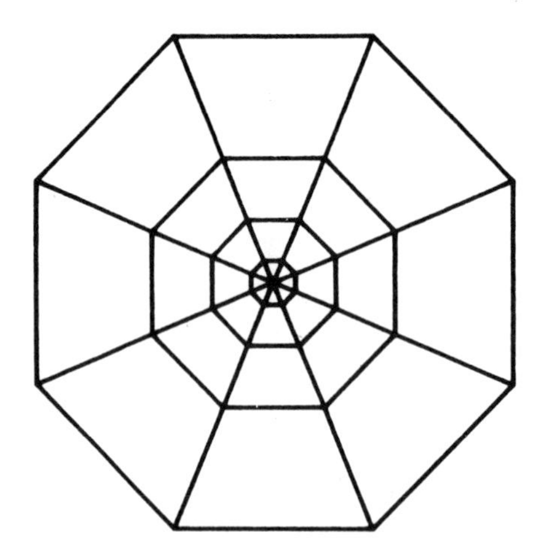

Bottom View

A–6. *Marigold.* This is an octagonal shape, usually made from flattish rough, which has a total of 73 facets (32 crown, 32 pavillion, 8 girdle, and the table). The main advantage of this cut is the ability to obtain a higher yield than would be possible with a conventional cut. In the very deep colors, especially brown and gold, this cut is very attractive. *(Information courtesy of Diamond Information Center)*

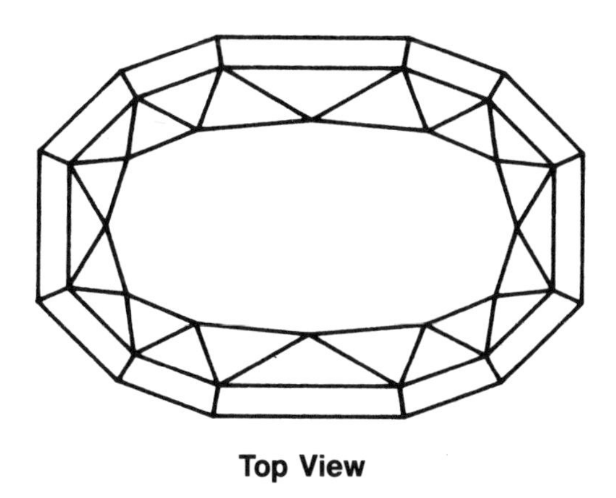

Top View

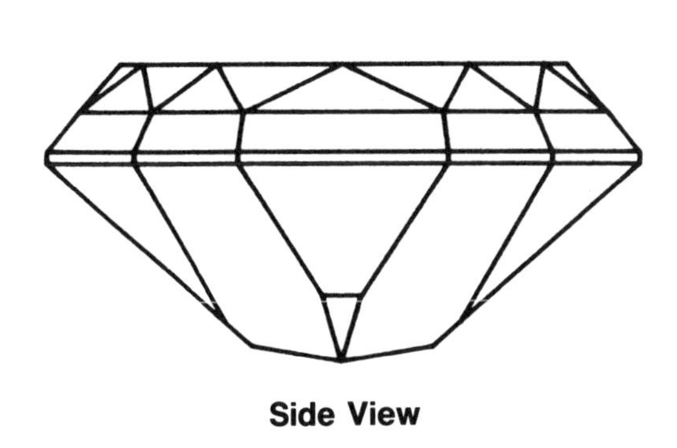

Side View

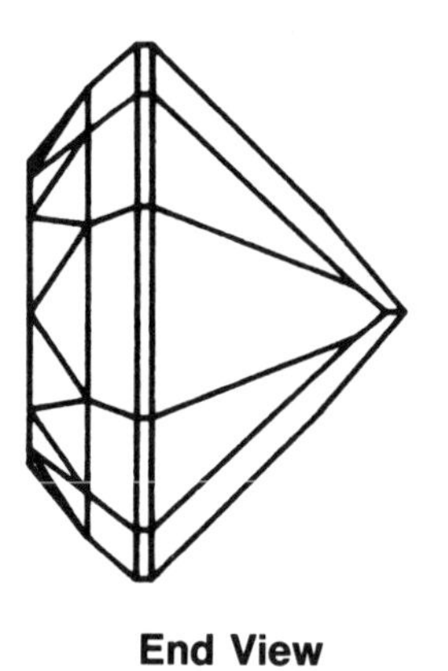

End View

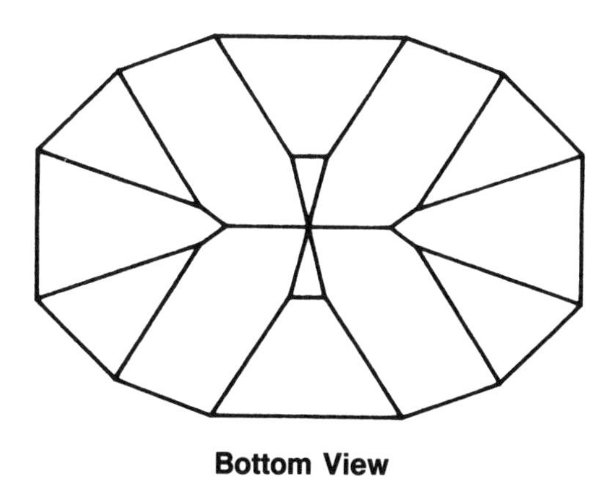

Bottom View

A–7. *Dahlia.* This is a twelve-sided oval shape. The principal advantages of this cut are higher yield over conventional shapes, producing an appealing stone in strong colors. *(Information courtesy of Diamond Information Center)*

Formulas for Unusual Diamond Cuts

Radiant or Barion Cut:

Length x Width x Depth x .0082

Barion Cut has 62 facets with an emerald-cut crown and round brilliant-cut pavilion.

Radiant Cut has 70 facets with a scissor-cut top and round brilliant-cut pavilion.

Trilliant Cut

Altitude x Base x Depth x .0057

Measure the narrowest width; this is referred to as the base. Then measure perpendicular to this to find the altitude. The trilliant-cut is shallow and requires a thicker girdle for durability. No length to width ratio is computed for this style. Adjust formula for thicker girdle between the point areas if present.

Rose-cut Diamond

Diameter^2 x Depth x 0.0043

Plus 10 to 25 percent for bulge factor (Formula courtesy of Mark Moeller, GG, CG).

To convert pennyweights into carats: *dwts. x 7.776 = carats*

Precious Metal Prices

Date	Gold	Silver	Platinum
1/82	$384.00	$ 8.03	$361.00
7/82	347.00	6.40	249.90
1/83	394.75	10.20	418.00
7/83	422.00	12.11	458.00
1/84	366.50	8.08	390.50
7/84	341.75	7.20	329.00
1/85	299.85	5.94	268.55
7/85	317.40	6.10	272.90
1/86	358.40	6.26	379.50
7/86	349.40	5.01	445.00
1/87	406.00	5.55	525.50
7/87	453.15	7.68	591.00
1/88	476.25	6.75	483.75
7/88	438.50	7.58	555.30
1/89	410.20	6.05	515.50
5/89	365.80	5.33	500.00

Source: The American Gem Market System

Average Trade Prices (per carat) The AGMS Network

GRADE*	AQUA G,FI	DIAM G,VVS2	EMER G,MI2	RUBY G,MI1	BSAP G,MI1	TOPZ G,FI	TOUR G,FI	TSAV G,LI2	TANZ G,FI
1/82	760	7250	5190	3550	1720	520	300	810	690
7/82	760	5090	4680	3550	1720	500	270	810	690
1/83	760	5610	4680	3550	1720	500	270	810	690
7/83	680	5500	4680	3590	1750	500	240	840	690
1/84	710	4900	4650	3300	1550	510	240	870	690
7/84	750	4600	4650	3275	1550	510	240	950	710
1/85	770	4600	4650	3225	1550	500	240	1025	820
7/85	760	4600	4650	3050	1425	500	240	1100	870
1/86	660	4500	3400	1950	1250	460	220	930	700
7/86	520	4000	3100	2050	1225	310	180	860	650
1/87	380	5550	2700	2100	970	210	170	850	580
7/87	380	5550	2925	1975	870	210	170	840	570
1/88	440	4650	2975	2100	900	200	160	710	610
7/88	430	4450	3000	2050	880	190	150	680	550
10/88	420	5550	3150	1900	820	200	150	630	540

Source: The American Gem Market System. All prices based on averages of actual transactions.
*Grades represent upper-medium quality gemstones.

Table of Comparative Ring-Size Equivalents

U.S.& Canada Standard	Average Diameter in Inches	British Equiv-alent	French Equiv-alent	German Equiv-alent	Japanese Equiv-alent	Swiss Equiv-alent	Inside Diameter in Milli-meters
000	.390	...	...	...	...	...	9.91
00	.422	...	...	...	...	...	10.72
0	.454	...	...	...	...	...	11.53
½	...	A	...	...	...	...	...
1	.487	B	...	...	1	...	12.37
1½	.503	C	...	...	...	...	12.78
2	.520	D	41½	13¼	2	1½	13.21
2½	.536	E	42¾	13¾	3	2¾	13.61
3	.553	F	44	14	4	4	14.05
...	...	G	45¼	...	5	5¼	...
3½	.569	...	...	14½	...	...	14.45
3¾	...	H	46½	...	6½	6½	...
4	.585	H½	...	15	7	...	14.86
4¼	...	I	47¾	...	...	7¾	...
4½	.601	I½	...	15¼	8	...	15.27
...	...	J	49	...	...	9	...
5	.618	J½	...	15¾	9	...	15.70
...	...	K	50	...	...	10	...
5½	.634	L	51¾	16	...	11¾	16.10
...	...	L½	...	...	11	...	...
6	.650	M	52¾	16½	12	12¾	16.51
6½	.666	N	54	17	13	14	16.92
7	.683	O	55¼	17¼	14	15¼	17.35
7½	.699	P	56½	17¾	15	16½	17.75
8	.716	Q	57¾	18	16	17¾	18.19
8½	.732	...	...	18½	17	...	18.59
...	...	R	59	...	...	19	...
9	.748	...	...	19	18	...	18.99
...	...	S	60¼	...	...	20¼	...
9½	.764	...	...	19½	19	...	19.41
...	...	T	61½	...	...	21½	...
10	.781	T½	...	20	20	...	19.84
10¼	...	U	62¾	...	21	22¾	...
10½	.797	U½	...	20¼	22	...	20.24
...	...	V	63¾	...	...	23¾	...
11	.814	V½	...	20¾	23	...	20.68
...	...	W	65	...	...	25	...
11½	.830	...	...	21	24	...	21.08
...	...	X	66¼	...	...	26¼	...
12	.846	Y	67½	21¼	25	27½	21.49
12½	.862	Z	68¾	21¾	26	28¾	21.89
13	.879	...	...	22	27	...	22.33

GUIDE TO GEMSTONE HANDLING

Explanation and notes on chart sections:

Section A: Hardness refers to the resistance of stones to being scratched. All steel tools may scratch stones with a hardness of 6 or less. The hard blue rubber wheels used to trim prongs during setting will scratch all stones except diamond. The softer gray pumice wheels should be used for colored stones above the hardness of 5. Use no abrasive wheels with stones under 5 in hardness.

Toughness refers to the danger of stones being damaged by the various processes used in the manufacturing and handling of stone set jewelry. A rating of excellent to good means that the stone will be safe if reasonable care is exercised. A rating of fair to poor means that special care must be taken or the process avoided entirely.

Section B: Setting . . .These caution points apply to all setting jobs.

1. Any stone with a knife edge girdle is dangerous to set. Pressure of prongs should be only above or below the girdle edge, never directly against it.
2. Examine stones carefully with a loupe for cracks or flaws on or near the points where prongs might be pressed. Avoid excess pressure in tightening such stones. If stones are too transluscent, use a pen light to help spot flaws.

Section C: Polishing involves the use of abrasive powders.

1. Rouge is a very fine abrasive and may be used with all stone jewelry except pearls.
2. Tripoli will scratch or dull the polish of stones with a hardness of 5 or less.
3. Stones listed as poor should be set after jewelry has been prepared and polished.

Section D: Repairs and Sizing:

1. Thorough cleaning of all jewelry needing repairs must be done.
2. Stones should be removed if repairs are required near the stone.
3. Stones which may be dyed or oiled will be damaged by such heat.
4. If repaires are made with stones set (as in sizing) be sure the stone has cooled before rinsing or pickling in acid. The center of the stone stays hot longer than the outside and time must be allowed for it to cool. If you must rinse while the piece is still hot, then do it in hot water. The rule is no fast temperature changes.
5. On sizing jobs with the stones set, the stone may be protected by keeping it upside down in a water filled bottle cap while soldering the bottom of the shank.

Section E: Boiling:

1. Never put a colored stone directly into boiling water. If the stone can take heat well, then put it in cold water and bring to a boil slowly.
2. Rinse in a hot water rinse if stone is still warm or in a warm water rinse if stone has cooled for a while. Again the rule is no fast temperature changes.
3. Do not boil rubies or emeralds. Many of these are "oiled" and they will lose color if boiled. Such stones should be removed before the jewelry is boiled and reset afterwards.
4. When cleaning such jewelry pieces, use a soft brush in lukewarm water with a mild soap or detergent.

Section F: Steaming:

1. Do not steam while stone is cold. Rinse first in warm water, not hot.
2. When steaming do not hold stone with tweezers. Hold only the mounting, not the stone.

Section G: Ultrasonic:

1. Solution should be kept warm if used as a cleaner after polishing.
2. Stones which have been heat treated are usually under strain and may crack in the ultrasonic process. Such stones are listed as poor and should be cleaned as listed under section E:4.

Section H: Acids:

1. Acids are used in the pickling and the plating processing. Again avoid fast temperature changes. Keep plating baths warm and rinse with warm water after using.
2. Do not use with stones which are organic or dyed. Stones listed as poor will be harmed by acids. They should be removed before repairing jewelry in which they are set.
3. Porous stones are also affected by acids. Examples are turpuoise, malachite, azurite, and shell cameos.

Reprinted courtesy of Howard Rubin and Leer Gem.

APPROXIMATE WEIGHT CHART FOR FACETED COLORED STONES

Weights are based upon the specific gravity of quartz (2.66).
Make appropriate changes for stones of higher or lower specific gravity.

OVAL	Approx. Wt.	OCTAGON	Approx. Wt.	ROUND	Approx. Wt.	PEAR	Approx. Wt.	NAVETTE	Approx. Wt.
5x3	.22	5x3	.25	4mm	.25	5x3	.21	6x3	.20
6x4	.40	6x4	.50	4½mm	.35	6x4	.40	8x4	.50
7x5	.80	7x5	1.10	5mm	.55	7x5	.70	10x5	1.00
8x6	1.20	8x6	1.50	5½mm	.70	8x5	.80	12x6	1.70
9x7	1.75	9x7	2.30	6mm	.90	9x6	1.35	14x7	2.50
10x8	2.50	10x8	3.00	6½mm	1.15	10x7	2.00	15x7	3.00
11x9	3.30	11x9	4.30	7mm	1.50	12x8	3.00	16x8	4.40
12x10	4.50	12x10	5.40	7½mm	1.70	13x9	3.65	18x9	7.00
14x10	5.50	14x10	6.60	8mm	2.05			20x10	11.00
14x12	7.00	14x12	8.40	8½mm	2.25				
16x12	8.50	16x12	10.75	9mm	3.00				
18x13	12.00	18x13	13.75	9½mm	3.20				
20x15	15.00	20x15	18.00	10mm	3.50				
				10½mm	3.75				
				11mm	4.00				
				11½mm	5.00				

Add or subtract 5 – 10 % for Heavy cut or Shallow cut stones.
Add or subtract appropriate %. factor for stones of higher or lower specific gravity.
These weights are an approximate estimation based on averages taken from parcels over a period of years. Varieties in cutting styles from different parts of the world will cause these averages to vary at times.
The weights given here are meant only as a guide and should not be used when exact appraisal requirements are needed.

STONE	HARDNESS & TOUGHNESS (read section A)	REACTION TO SETTING (read section B)	REACTION TO POLISHING (read section C)	REACTION TO SIZING AND REPAIRS WHICH REQUIRE TORCH (read section D)	REACTION TO BOILING (read section E)
DIAMOND	H. 10 T. good	very good	excellent	good	excellent
RUBY AND SAPPHIRE (Corundum)	H. 9 T. very good	very good	excellent	Ruby-good Sapphires may lose color when heated	good
CATSEYE & ALEXANDRITE (chrysoberyl)	H. 8½ T. very good	very good	excellent	good-fair; remove if repairs are made near stone	good
SPINEL	H. 8 T. good-fair	very good-fair	very good	good-fair; remove if repairs are made near stone	good-fair
PRECIOUS TOPAZ	H. 8 T. poor	fair-poor; take care stone cleaves easily	good	poor; stones may crack or lose color	poor
EMERALD (beryl)	H. 7½-8 T. poor	poor; stones usually flawed and under strain	fair; do not apply heavy pressure	poor; stones should never be heated	Poor; should be cleaned in lukewarm water only.
AQUAMARINE (beryl)	H. 7½-8 T. good-fair	good-fair	good	poor; stone may change color with heat	fair-poor; avoid fast temperature changes.
TOURMALINE	H. 7-7½ T. good-fair	good-fair	good	fair-poor	fair
GARNET incl. RHODOLITE & TSAVORITE	H. 6½-7½ T. good-fair	good-fair; flawed stones are under strain	good	fair-poor; play safe, remove expensive stone before repair	fair-poor
RUTILE & FABULITE (Synthetic)	H. 6½-7 T. poor-very poor	very poor; will take very little pressure	very poor; use very light pressure or set after polishing	very poor; stones will crack with heat	poor; stone may crack
AMETHYST & CITRINE (quartz)	H. 6½-7 T. good	good	good	fair; color may change with heat	fair
PERIDOT	H. 6½-7 T. poor-very poor	poor; facet edges chip easily	poor	very poor; remove stone before repairs or sizing are made	poor; avoid extreme temperatures
TANZANITE (zoisite)	H. 6½ T. poor	poor	fair (avoid heavy pressure)	very poor; remove before repairs are made	poor
JADEITE & NEPHRITE (jade)	H. 6-7 T. Excel.	excellent	fair; tripoli may damage polish on stone. Use only rouge	Poor, no repairs near stone	Good; heat may discolor dyed material
KUNZITE & HIDDENITE (spodumene)	H. 6-7 T. very poor	poor	fair	Poor; stones will lose color	Poor; may crack if boiled
ZIRCON	H. 6-6½ T. poor	poor	fair	poor	poor
MOONSTONE (feldspar)	H. 6-6½ T. fair-poor	good-fair	good-fair	poor	poor
OPAL – Also doublets & triplets	H. 5½-6½ T. very poor	poor	poor (avoid heavy pressure)	very poor; remove before repairs are made	Poor; boiling will crack stone, triplets separate
HEMATITE	H. 5-6½ T. good-fair	good-fair	good-fair	poor	good
TURQUOISE	H. 5-6 T. good-poor	fair	fair	very poor; stone will explode with heat	poor; may lose color
LAPIS LAZULI (lazurite)	H. 5-6 T. fair-poor	fair	fair-poor; tripoli will harm polish on stone	poor	fair-poor; some dyed stones will lose color
SHELL CAMEO	H. 3½ T. poor	poor, will crack with excess pressure	poor; polish jewelry lightly with very little pressure	Cannot take heat of repair. Will show burn marks	Color will fade if boiled
CORAL	H. 3-4 T. good-poor	good	poor; use rouge only	very poor; remove stone before repair	poor; may lose color
PEARLS & MOBES	H. 2½-4½ T. fair-poor	fair; mobes take pressure poorly	poor; will affect luster badly	poor; pearls will burn	poor; will lose color tint. Mobes separate
IVORY	H. 2½-3 T. fair	fair	fair-poor; use light pressure	poor; heat will cause stone to shrink	fair-poor; dyed pieces may lose color
AMBER	H. 2-2½ T. poor	very poor; will scratch easily	poor	very poor; stone will melt or burn	very poor; do not boil

Reprinted courtesy of Howard Rubin and Leer Gem.

REACTION TO STEAMING (read section F)	REACTION TO ULTRASONIC (read section G)	REACTION TO ACIDS PICKLING & PLATING (read section H)	COMMENTS (add your own comments based upon your own experience)
excellent	excellent	excellent	
good	good	good	Watch for oiled stones. Do not heat
good	good	good	
good	good	good	
poor	fair	good	Any heating may discolor or crack stone.
poor	fair	poor; stones may crack or lose oil if solutions are too hot	Avoid all heat; Chatham and Gilson synthetics react the same as natural stones.
fair	fair	good	
fair	good	good-fair	May change color with heat during repairs
fair	good	fair-poor; acids may affect polish on stone	
poor	fair-poor	fair	Reacts poorly under heat and pressure
fair	good	good-fair	
fair-poor	fair	poor	Should not get much heat or pressure
poor	poor	fair	Will not take much heat or pressure
good	good	poor; acid will affect polish on stone	
poor	fair	fair	Heat may fade color
poor	fair	fair	Does not take heat well
fair	fair	fair-poor	
poor	fair-poor	poor	Opals should be examined by shining a light through the stone to see if there are cracks. Do not process cracked stones.
good	good	poor; acids attack stones	
fair	fair-poor	very poor; will dissolve in acids	Takes heat and pressure poorly. Color may fade in untreated stones.
good	good-fair	poor; will change color. Acid will attack Pyrite & Calcite inclusions	Many lapis are dyed. Colors may change with heat or acids.
fair-poor	fair	very poor; will dissolve in acids	Cameos made of shell are very delicate and will not take much heat or pressure.
fair	fair	very poor; will dissolve in acids	Much coral is dyed and will be affected by heat.
fair	fair	very poor; will dissolve in acids	Watch for spot in nacre which may be hollow underneath
good	fair	fair	Many imitations available, all react differently.
poor-fair	poor	very poor; will dissolve in acids	Many imitations will react the same

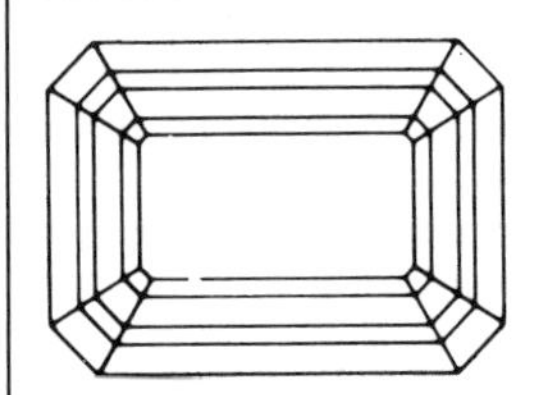

COLORED STONE GRADING REPORT

OF COLOR AND QUALITY ANALYSIS

issued by

An independent gemological laboratory for the grading of diamonds and colored gemstones.

Laboratory Report:

Date:

This grading report has been prepared by a G.I.A. Graduate Gemologist. and/or F.G.A. Fellow of the Gemological Association of Great Britian and Accredited Gemologists Association Member.

Public use of any part or all of the contents of this report is absolutely prohibited.

NOTE: This report contains a description of the characteristics of a colored stone based on the application of accepted gemological grading techniques

This report is issued at the request of the client submitting the above mentioned registered stone and is for his exclusive use only.

SPECIES:

VARIETY:

Shape and Cut:

Carat Weight:

Measurements:

COLOR GRADE:

Primary Hue:

Secondary Hue(s):

Color/Scan Notation:

ColorMaster Notation:

TONAL GRADE:

LIGHT SOURCE:

CLARITY GRADE:

CUTTING GRADE:

Proportions:

Depth:

Finish:

Symmetry:

BRILLANCY GRADE:

COMMENTS:

F.G.A., G.G.

Gemologist

UNLESS OTHERWISE STATED, ALL COLORED STONES LISTED ON THIS GRADING REPORT HAVE PROBABLY BEEN SUBJECTED TO A STABLE AND POSSIBLY UNDETECTABLE COLOR ENHANCEMENT PROCESS. PREVAILING MARKET VALUES ARE BASED ON THESE UNIVERSALLY PRACTICED AND ACCEPTED PROCESSES BY THE GEM AND JEWELRY TRADE.

IMPORTANT: THIS GRADING REPORT REPRESENTS AN ESSENTIAL ELEMENT IN THE OVERALL EVALUATION OF THE SUBJECT GEMSTONE WHICH MUST BE REFLECTED IN A COMPLETE APPRAISAL BY A GRADUATE GEMOLOGIST/APPRAISER AND IS NOT INTENDED TO BE AN INDICATION OF VALUE UNLESS INCORPORATED IN A COMPLETE APPRAISAL FORMAT.

Form courtesy of Tenhagen Gemstones.

* All color grading analysis is subject to the color temperature of the light source and the grader's sensitivity to color.
* The commercial desirability and the quality of any color analysis is based on the interrelationships of all the factors indicated under the color grading system. Conclusions may vary due to the subjectivity of colored stone analysis.

COLOR GRADE (AGL)

1 2	3 4	5 6	7 8	9 10
EXCELLENT	VERY GOOD	GOOD	FAIR	POOR

The color grade is the overall color perceived by a combination of primary and secondary hues.

TONAL GRADE (AGL)

0 5 10 15 20	25 30 35	40 45 50	55 60 65	70 75 80 85 90	95 100
VERY LIGHT	LIGHT	LIGHT-MED.	MEDIUM	MEDIUM DARK	DARK-V. DARK

The Tonal Grade is a gemstone saturation on a light to dark scale where 0 = colorless and 100 = black.

CLARITY GRADE (AGL)

FL	LI1 LI2	MI1 MI2	HI1 HI2	EI1 EI2 EI3
Free of Inclusions	Lightly Included	Moderately Included	Heavily Included	Excessively Included

The Clarity Grade is determined by number, nature, size and relative location of inclusions to the unaided eye and 10x magnification.

CUTTING GRADE

1	2 3	4 5 6	7 8	9 10
EXCELLENT	VERY GOOD	GOOD	FAIR	POOR

The Cutting Grade is determined by the proportions, depth, finish and symmetry.

BRILLIANCY GRADE

100%	90-80%	70-50%	40-30%	20-10%
EXCELLENT	VERY GOOD	GOOD	FAIR	POOR

Brilliancy Grade is the percentage of internal reflected light that is returned to the eye through the crown.

GEMOLOGICAL INSTRUMENTS AND EQUIPMENT USED

[1] Nikon SMZ-10 Microscope with dark field illumination
[2] Deluxe "B" Gemolite
[3] Eickhorst Gemmaster Immersion Base with Nikon SMZ-10
[4] Gem Diamondlite
[5] Eickhorst Dialite
[6] Duro-Test: Vita Lite
[7] Duplex II Refractometer
[8] Anderson/Payne Spinel Refractometer
[9] Standard Rayner Refractometer
[10] Rayner Dialdex Refractometer
[11] S+T CubiCz Refractometer
[12] Kruss/Riplus Refractometer
[13] Rayner Sodium Light Source
[14] Polariscope
[15] Calcite Dichroscope
[16] Spectroscope Unit
[17] Master Color Diamonds
[18] Eickhorst Diamond-Photometer
[19] ColorMaster
[20] Color/Scan
[21] Gem Color Guide
[22] Gem Dialogue
[23] Proportion Scope
[24] Mettler Scale PC 400C
[25] Scientech SE300 CT/DWT/Grams
[26] Ceres Diamond Probe
[27] U.V.-254/366 Mineralight Lamp
[28] Micron Millimeter Micrometer
[29] Leveridge Gauge
[30] Micro 2000 Digital Micrometer
[31] Specific Gravity Liquids
[32] Thermal Reaction Tester
[33] Metal Test Kit
[34] Polaroid CU-5
[35] 35 mm Camera
[36] Pinpoint Illuminator
[37] Fibre Optic Illuminator
[38] Geiger Counter

CUBIC ZIRCONIA STONE CONVERSION CHART

ROUNDS

mm	Diamond Size	Approx. Cz Wt.	mm	Diamond Size	Approx. Cz Wt.
1.50	.02	.03	3.75	.20	.35
1.75	.025	.04	4.00	.25	.45
2.00	.03	.05	5.25	.50	1.00
2.25	.04	.07	6.00	.75	1.45
2.50	.05	.10	6.50	1.00	1.80
2.75	.07	.14	7.50	1.50	2.85
3.00	.10	.19	8.00	2.00	3.50
3.25	.13	.25	9.00	2.50	4.70
3.50	.16	.30	9.50	3.00	5.65

FANCIES

Marquis	Pear	Oval	Heart	Emerald	Dia. Size	Cz Wt.
6x3	5x3	5x3	4x4	5x3	.25	.50
7x3.5	6x4	6x4	5x5	6x4	.50	.75
10x5	8x5	7.5x5.5	6.5x6.5	7x5	1.00	1.75
11x5.5	9x6	8.5x6.5	7x7	8x6	1.50	2.75
12x6	10x7	9x7	8x8	8.5x6.5	2.00	3.25
13x6.5	11x7.5	9.5x7.5	8.5x8.5	9x6.5	2.50	4.25
14x7	12x8	10x8	9x9	9x7	3.00	5.25

MAJOR COMPARISONS OF CUBIC ZIRCONIA TO DIAMONDS

Specification	Cubic Zirconia	Diamond
Hardness	8.75	10.00
Specific Gravity	5.95	3.52
Refraction	2.15	2.40
Dispersion	0.06	0.04

Birth Stones Through the Ages

Month	Hebrews	Romans	635. A.D. Idisorus, Bishop of Seville	Arabians	Poles	Italians	18th to 20th Centuries	Encyclopedia Brittanica "Current Acceptance"	Webster's Unabridged Dictionary	Present Popular List	Synthetic Stones
Jan	Garnet	Garnet	Hyacinth	Garnet	Garnet	Jacinth Garnet	Hyacinth Garnet	Garnet	Garnet	Garnet	Garnet
Feb	Amethyst	Amethyst	Amethyst	Amethyst	Amethyst	Amethyst	Amethyst Hyacinth Pearl	Amethyst	Amethyst	Amethyst	Amethyst
March	Jasper	Bloodstone	Jasper	Bloodstone	Bloodstone	Jasper	Jasper Bloodstone	Bloodstone	Jasper Bloodstone	Bloodstone Aquamarine	Aquamarine
April	Sapphire	Sapphire	Sapphire	Sapphire	Diamond	Sapphire	Diamond Sapphire	Diamond	Diamond Sapphire	Diamond	White
May	Chalcedony Carnelian Agate	Agate	Agate	Emerald	Emerald	Agate	Emerald Agate	Emerald	Emerald	Emerald	Emerald
June	Emerald	Emerald	Emerald	Agate Chalcedony Pearl	Agate Chalcedony	Emerald	Emerald Agate Catseye Turquoise	Pearl	Agate	Pearl Moonstone	Alexandrite
July	Onyx	Onyx	Onyx	Camelian	Ruby	Onyx	Turquoise Onyx, Ruby	Ruby	Turquoise	Ruby	Ruby
Aug	Carnelian	Carnelian	Carnelian	Sardonyx	Sardonyx	Carnelian	Sardonyx Carnelian Moonstone Topaz	Sardonyx	Carnelian	Sardonyx Peridot	Peridot
Sept	Chrysolite	Sardonyx	Chrysolite	Sardonyx	Sardonyx	Chrysolite	Beryl Chrysolite	Sapphire	Chrysolite	Sapphire	Sapphire
Oct	Aquamarine Beryl	Aquamarine Beryl	Aquamarine Beryl	Aquamarine Beryl	Aquamarine Beryl	Beryl	Aquamarine Beryl, Pearl Opal	Opal	Beryl	Opal Tourmaline	Rose
Nov	Topaz	Topaz	Topaz	Topaz	Topaz	Topaz	Topaz Pearl, Opal	Topaz	Topaz	Topaz Citrine	Golden Topaz
Dec	Ruby	Ruby	Ruby	Ruby	Turquoise	Ruby	Ruby Bloodstone	Turquoise	Ruby	Turquoise Lapis Lazuli	Blue

Manufacturers and Wholesalers of Antique Reproduction Jewelry

The following list is provided as a resource for catalogs and information on antique jewelry reproductions. This is not a comprehensive list and does not include those Pacific-rim countries that manufacture period reproductions.

Herzog and Adams
37 West 47th St.
New York, NY 10036
Complete line of reproduction jewelry with garnets and marcasite. Catalog is available.

K. Goldschmidt
Jewelers, Inc.
24 West 47th St.
New York, NY 10036
Well crafted 14K items with natural gemstones. Complete line of rings, necklaces, earrings, pendants, brooches, and slide bracelets.

Heirloom 73
22 Throckmorton St.
Freehold, NJ 07728
The company has three divisions: *Heirloom 73* – 1920s style filigree 14K jewelry; also vermeil and sterling jewelry; *Vintage Creations* – Sterling silver Art Deco style jewelry; *Legacy* – Gold filigree jewelry in 18K.

Judith Jack, Inc.
27 West 47th St.
New York, NY 10036
Art Deco style jewelry in sterling and marcasite. Also gold articles in antique style replica goods.

Jewelry by Joshua
157 N.E. 166 St.
Miami Beach, FL 33162
Reproduction slides and slide bracelets.

Bernard Nacht & Co., Inc.
29 West 47th St.
New York, NY 10036
Catalog is available of antique and replica jewelry.

Vintage Jewelers
P.O. Box 342
Monsey, NY 10952
Reproductions of 14K Victorian-style jewelry.

Auction Galleries That Hold Jewelry Sales

Christie's
502 Park Avenue
New York, NY 10022

Butterfield & Butterfield
220 San Bruno Avenue
San Francisco, CA 94103

William Doyle Galleries
175 East 87th St.
New York, NY 10128

Leslie Hindman Auctioneers
215 West Ohio St.
Chicago, IL 60610

Phillips Ltd.
406 East 79th St.
New York, NY 10021

Robert W. Skinner Galleries
#2 Newbury St.
Boston, MA 02116

Sotheby's
1334 York Ave.
New York, NY 10021

A. Weschler & Son
905–9 E. St. NW
Washington, DC 20004

Wolf Auction Gallery
13015 Larchmere Blvd.
Shaker Heights, OH 44120

Robert C. Eldred, Inc.
1483 Main St.
Route 6A
East Dennis, MA 02641

Dunning's Auction Service, Inc.
755 Church Road
Elgin, IL 60123

Auction Newsletter

For a monthly newsletter (by subscription) with in-depth information on 20 to 30 regional auction markets and final recorded auction results:

Auction Forum Limited
341 West 12th St.
New York, NY 10014

Pearl Facts

Pearl Composition

Carbonate of Lime (aragonite)	S.G. = 2.94
Conchiolin.	S.G. = 1.34
Water.	S.G. = 1.00

Mollusk Name	Mollusk Locale	Color of Pearl	Specific Gravity
Margaritifera Vulgaris	Persian Gulf (Arabian coast)	Creamy-white	2.68 to 2.74
Margaritifera Vulgaris	Gulf of Manaar (Sri Lanka coast)	Pale cream-white	2.68 to 2.74
Margaritifera Margaritifera	North coast of Australia	Silver-white	2.68 to 2.78
Margaritifera maxima	Northwest coast of Australia	Silver-white	2.67 to 2.78
Margaritifera carcharium	Sharks Bay, W. Australia	Yellow	2.67 to 2.78
Margaritifera radiata	Venezuela	White	2.63 to 2.75
Margaritifera martensi	Japan (natural)	White, with greenish tinge	2.66 to 2.76
Strombus gigas (the great conch)	Bahamas and Florida	Pink	2.85
Haliotidœ (the abalone)	West coasts of Mexico and California	Green, yellow, blue, black	2.61 to 2.69
	Gulf of Mexico	Black	2.61 to 2.69
Unio	**Freshwater pearls** North American	White	2.66 to over 2.78
Unio margaritifera	Europe	White	2.66 to over 2.78
Margaritifera martensi	**Cultured Pearls** Japan	White	2.72 to 2.78

(Source: Robert Webster, FGA)

Size of Pearl	Number of Pearls in a Momme
3 mm	90 pearls
4 mm	40 pearls
5 mm	20 pearls
6 mm	12 pearls
7 mm	7 pearls
8 mm	5 pearls
9 mm	3 pearls

The size of natural pearls is sometimes given in "pearl grains." There are 75 pearl grains to a momme. One momme equals 18.75 carats.

Natural Pearl Conversions

Indian

A rati is equal to 0.91 carat.
A tola is equal to 58.18 carats (by physical weighing of a known weight).
A magelin is equal to 1.75 carats.
64 ratis equal 1 tola.
To convert ratis to carats multiply by 10 and divide by 11.

Burmese

A rati is equal to 0.91 carat.
A bali is equal to 58.18 carats (64 ratis).
A tickal is equal to 80 carats (88 ratis).
A viss is equal to 880 carats.
A lathi is equal to 1.75 carats.

Ceylonese

A chevvu, chow, or tank is equal to 21.84 carats (24 ratis).
A manchadi is equal to 1.15 carats

Thai (Siamese)

A catty is equal to approximately 3,015 carats.

Iranian (Persian)

A miscal is equal to 36.40 carats (40 ratis).
2 miscals are equal to 1 dirhem.

Turkish

A checky is equal to 1600 carats (320 grammes).

Brazilian

An oitava (octavo) is equal to approximately 17.5 carats.

Courtesy of The Guide, Gemworld International, Inc.

WORKING AGREEMENT

APPRAISER'S LETTERHEAD

Dear

This letter will confirm our agreement as to the preparation of the appraisal to be done on ________________.

The property to be appraised is broadly described as follows:

The property is located at ________________ where you will arrange for us to physically inspect the property and at which time you will supply us with any factual information within your knowledge we may request from you. In preparing the appraisal, ________________ will inspect the property, prepare a detailed description of the property, evaluate it and submit a written and signed report to you. The appraisal will be prepared for the purpose of ________________ and will be used by you only for such purpose.

The fee for said appraisal shall be ________________ per hour/ day. In addition, expenses for traveling and other out-of-pocket expenses shall be charged to you. The fee and expenses shall be due and payable upon delivery of the written report. An advance of ____% of the estimated fee and/or expenses is due (upon signing this letter of agreement) (upon commencement of the appraisal inspection). An additional fee will be charged for any required future services pertaining to this appraisal.

Our appraisal will represent our best judgment and opinion as to the current fair market value and other factors stated in the appraisal of the appraised property. However, the appraisal will not be statement or representation of fact nor is it a representation or warranty with respect to authenticity, genuineness, or provenance.

Any controversy or claim arising out of or relating to this contract, or the breach thereof, shall be settled by arbitration in accordance with the Commercial Arbitration Rules of the American Arbitration Association, and judgment upon the award rendered by the Arbitrator(s) may be entered in any Court having jurisdiction thereof.

If you are in agreement with these terms, please sign and return one copy of this letter to us at your earliest convenience.

Looking forward to working with you, we are,

Very truly yours,

________________	________________
Appraiser	Client
________________	________________
Date	Date

Reprinted Courtesy of the Appraisers Association of America

AFFIDAVIT PAGE FOR ESTATES

APPRAISER'S LETTERHEAD

IN THE MATTER OF THE
ESTATE TAX
UPON THE ESTATE OF

_______________________,
Deceased

STATE OF)
: SS
COUNTY OF)

(NAME OF APPRAISER), being duly sworn, deposes and says:

I am an appraiser of (TYPE OF PROPERTY) including that of the kind and character set down upon the annexed schedule of (NUMBER OF PAGES). A resume of my qualifications is attached to this appraisal).

On (DATE), I personally examined, inventoried, and appraised the fair market value of certain tangible personal (TYPE OF PROPERTY) said to belong to the Estate of (NAME OF DECEASED), on the premises at (LOCATION). A schedule of such property, consisting of _______ pages and made a part of this affidavit, is hereto annexed. The appraised fair market valuation for estate purposes for each article or group of articles is set opposite the description of the same as of the date of death of said decedent, which is (DATE OF DEATH). The total fair market value of such property as of the date of death is (DOLLARS), such fair market value being to the best of my knowledge and belief that which would have been paid for such property by a willing buyer to a willing seller in an open market and not as a forced sale.

I state that my fee for this appraisal is not contingent upon the stated valuations, and that I have no present or future contemplated interest in any of the appraised articles.

Sworn to, before me this

day of , 19

APPRAISER

____
Notary Public

Reprinted Courtesy of the Appraisers Association of America

GLOSSARY

Appraisal Terms

Aesthetic value. Regarded as the qualities of design, form, shape, structure, texture, decoration, and color that rest in the article and can be described. This is a value appreciated differently in different cultures. A value of an object in relation to its beauty and artistic merit.

Appraisal. An estimate or opinion of value.

Appraise. To express an opinion as to some form of property, real, personal, or intangible. To determine the value of property after investigation, collection and analysis of pertinent data. To exercise objective and subjective judgment of the worth of property. To value property.

Appraisal date. May also be called the report date (day report is written). It is not necessarily the *date of value* which is the date of the last examination of the subject property.

Appraisal process. A systematic analysis of the factors that bear upon the value. An orderly program by which the problem is defined, the work necessary to solve the problem is planned, the data involved acquired, classified, analyzed and interpreted into an estimate of value.

Appraiser. One who estimates value; specifically, one who possesses the necessary qualifications, ability, and experience to execute or direct the appraisal of real or personal property.

Attribute. Any property, quality, or characteristic that can be ascribed to a person or thing.

Authenticate. To render authentic or valid. To determine as genuine, to prove the authenticity of, to verify.

Certificate. Legal document giving formal assurance of the existence of a fact or set of facts. It is an assurance within a known body of knowledge.

Comparables. Like kind or identical items to that which is being appraised; used for comparison in identification and property sales.

Cost. The monetary figure paid for an item; the price. The amount actually spent in producing or manufacturing an article. A figure paid by a wholesale dealer or retailer to which a markup is added for consumer selling price.

Data. Information pertinent to a specific assignment. The data may be general and specific.

Depreciation. A loss of value caused by loss of utility. Depreciation may be caused by deterioration and/or obsolescence. Physical depreciation, deterioration, is marked by wear and tear, decay and/or defects. Obsolescence can be of two types: economic or functional. Economic obsolescence is caused by external factors to a property such as style changes. Functional obsolescence has to do with changes within the property itself.

Disclaimer. Denial or disavowal of an interest in, or claim to, any interest in an estate or appraisal. Also, may be a statement of limiting conditions in an appraisal under which the appraisal was performed and subject.

Evaluation. As distinct from valuation, an evaluation report provides advice or counseling to a client regarding the client's problem and proposed alternative solutions to that problem.

Expert. An expert is one who, by reason of education, experience, or study, is presumed to have special knowledge of, or skill in, a particular field.

Fair Market Value. A legal term. Value that is reasonable, unprejudiced, realistic, and consistent with known facts. The price at which the property would change hands between a willing buyer and a willing seller, neither being under any compulsion to buy or sell, both having reasonable knowledge of relevant facts.

Function of Appraisal. The reason for which the appraisal is produced or its intended use, for example, insurance, divorce, taxes.

Highest and best use. The utilization of a property to its best and most profitable use. The use is chosen from among reasonably probable and legal alternative uses, which is found to be physically possible, appropriately supported, and financially feasible to result in its highest value. In jewelry, it is applied as the "most appropriate marketplace."

Identification. The principle of identification specifies uniquely the character and quality of a property.

Intrinsic value. The amount of money which is equivalent to the inherent worth in a thing, for example, the gold in a 14K gold item.

Limiting condition. A statement in the appraisal that the appraisal is subject to and contingent upon certain conditions of examination or outside forces.

Market price. The money actually paid for an article, or the figure asked for an item. It differs from market value because it is an accomplished or historical fact, where market value remains an estimate until accomplished.

Market value. Economic concept, the most probable price in terms of money which a property will bring in competitive and open market with all conditions requisite to a fair sale, the buyer and seller acting prudently and knowledgeably and assuming the price is not affected by undue stimulus.

Opinion. Interpretation and evaluation.

Price. The quantity of one thing which is exchanged for another; the amount of money paid, asked, or offered where sale is contemplated; the money expected or given in exchange for commodities or services. There is a distinction between price and value. The price of an item may be more or less than its value.

Provenance. Origin and history of ownership.

Purpose. The type of value being estimated in an appraisal, such as retail replacement, fair market value, liquidation, scrap, and so forth.

Valuation. The process of estimating value on an item.

Value. The quantity of one thing which can be exchanged for another, projected justified price estimate.

Auction Catalog Terms

After. In the opinion of the auction house, a copy of the work of the artist.

Antique. Jewelry that is one hundred years or older.

Archaeological jewelry. Usually that of the 19th century inspired by Etruscan finds.

Art Deco. Abstract designs and geometric patterns characterize the jewelry.

Art Nouveau. Jewelry with free flowing and curved lines created from 1890 to the early 1900s.

Attributed to. Means the work is of the period of the named artisan and it may be, in part or entirely, his work.

Baguettes. Stones cut in long narrow rectangles.

Brilliant-cut. Modern full-cut round diamonds or other gems cut with 58 facets.

Bears signature. Has a signature which in the qualified opinion of the auction house might be the signature of the artist.

Calibre. Stones cut in a shape to fit snugly with other stones in a setting.

Cannetile. Thin wires used to make a filigree design or lacy-work pattern.

Circle of. In the opinion of the auction house, the item is a work of the era, either closely associated with the artist or from his studio.

Dog collar. A wide collar necklace, usually multiple strands of pearls or beads, that fit snugly around the neck.

Dated. The item is dated and, in the opinion of the auction house, it was completed at about that date.

En tremblant. A piece of jewelry constructed so that it moves like a feather swaying in the breeze. The aigrette was typical of *en tremblant* jewelry.

Foiled. Colored foil placed behind or beneath a stone to enhance the stone's color.

Granulation. A technique of small beads of gold forming designs on a piece of jewelry.

Hunting case watch. A watch that has a cover.

Manner of. In the opinion of the auction house, the jewelry is in the style of the artist, but possibly at a later period.

Mourning jewelry. Jewelry worn to remember a loved one or one deceased. Dark materials such as jet were used, often hair of a loved one.

Parure. A suite of jewelry comprised of a necklace, bracelet, earrings, brooch, and possibly a diadem. A demiparure would have less matching pieces, possibly two.

Rose cut. A style of diamond cutting with a flat base and a dome-shaped top covered with a varied number of triangular facets, and terminates in a point. It is an early style of cutting thought to have originated in India. There are three-facet, six-facet, and full Dutch rose cuts.

Sautoir. Long neck chain usually 36 to 60" long and hanging below the waist.

Signed. Signatures, stamped or hallmarked from a particular firm or individual artist.

School of. A work by a pupil or follower of the artist and executed in the style of the master.

Studio of. A work possibly completed by a pupil with supervision of the artist (master).

Style. Produced to look like jewelry of an earlier period (in the style of . . .).

Workshop of. Possibly a work finished in the shop and under the supervision of a well-known artist.

Enamel Terms*

Basse-tailé. Meaning low cut, is a technique where the metal groundplate is chased, shaped, or engraved in such a manner that the surface design beneath the fused enamel forms an essential design element in lightening and darkening the enamel by depth (origin: Italy 1400).

Camaieu. Is a limoges-type method where an opaque white or grey enamel, often depicting a figure or object, is painted atop a transparent background (see grisaille).

Champlevé. A technique where the enamelling is fused into depressed areas which are chased, stamped or engraved. Basse taille is an additional method to champleve. (Celt c. 200 B.C.)

Cloisonné. The earliest recorded technique where the enamel powder is placed within metal gated cells before firing. The high heat fusion binds the metal ribbon wall to the enamel and to the metal groundplate (Greek c. 500 B.C.).

Eglomise. A technique of decorating post-fired enamels with the application of gold-leaf which is then point-scribed to expose the under leaf enamelling (France c. 1700).

Email de riz. Also called pate de riz, an opalescent greyish enamelling (France c. 1700).

Email de verre. Is a multi colored (polychrome) enamel without champleve depressions or cloisons. The patches of color slightly blend at their edge. Seen in some molded glass objects (France c.1600).

En blanc. A translucent/opaque white enamelling.

En plein. Technique where the enamelled surface is broad and of one color and type atop a polished/ engraved base.

En plein sur fond reservé. A form of en plein enamelling but restricted in area. Common in the work of the Czarist Russian enamellists (France c. 1750).

En resillé sur verre. Meaning 'in a network on glass', this technique applies enamel atop glass or rock crystal quartz and is often foil backed.

En ronde bosse. A three-dimensional application where the enamelled field is fused atop an irregular base. In this method, the supports needed during firing add to the difficulty. Up to ten firings may be needed, due to the dripping which may occur. Cane handles are typical of this method.

Encrusting. The same as enamelling en ronde bosse.

(Reprinted courtesy of Tom R. Paradise, T. R. Paradise & Co.)

Flinking. The groundplate decoration composed of parallel and concentric engraved bands and lines seen beneath transparent enamels. Common with *en plein* technique (Byzantine c. 1400).

Fuming. A post-firing technique where a metal chloride is applied atop the enamel upon immediate removal from the kiln. Often an iridescent or metallic shimmer is attained.

Grisaille. A technique in which white opaque/translucent enamels are painted atop a black enamelled base (France c.1700).

Groundplate. The base or background upon which the enamel is attached by fusing, metal attachment, or applique.

Impasto. A coarse method in which enamel powder is applied to the groundplate in a spontaneous manner and fired very quickly, often less than 75 seconds. This has a rough and irregular look, bumpy and inconsistent in transparency (twentieth century).

Lattimo. See *verre de lait.*

Limoges. A painting technique on enamel that involves various layers of fine painting-grade enamels fired after each application to produce a picturelike image without cloisons (France c.1400).

Moire. A groundplate engraved design depicting numerous concentric wavy bands/lines.

Plique-a-jour. A technique of producing cloisonne-type enamelling without any metal groundplate, attaining a stained glass effect. Common to the pieces of Art Nouveau period when comprised of transparent/translucent enamels.

Plique-a-nuit. The same as *plique-a-jour* but using an opaque/translucent black/dark brown enamel.

Quilloché. A metal groundplate decoration motif depicting two opposed, meandering serpentine shapes. This is often a misnomer used in describing the transparent/translucent enamels used atop such background decorations. Often seen utilized by the Czarist Russian enamellists, used *en plein.*

Rocaille. A technique of layering in a thick manner opaque brown and black enamels, producing a rocklike look.

Sgraffito. A post-firing technique in which a sharp tool is used to engrave the enamelled surface to expose some metal background through the enamelling. The exposed metal displays a fine hairlike shine through the enamel (Italian c.1400).

Sunburst. A groundplate engraved motif displaying straight but radially emanating lines or bands; seen often in framing.

Sunray. A groundplate engraved motif displaying waving but concentric and radially emanating lines or bands (*see* Sunburst).

Taille d'epargne. Meaning a saving cut, this technique involves the delicate engraving of thin lines and the subsequent application of an opaque black enamelling. These narrow black lines are often seen in middle to late Victorian pieces as fine black accents.

Verre de lait. An enamelling of opaque white often in the *en plein* manner. Imitates milk, hence the name. An enamelling technique often used by Fabergé.

Pearl Terms

Pearl. An object that originates in the body of a mollusk. The pearl consists of concentric layers of the mineral argonite with conchiolin resulting in the *nacre.* The term "pearl" used alone refers to a natural formation. FTC considers it unfair practice to use the term when referring to a cultured pearl unless the word "cultured" or "cultivated" precedes the word "pearl." Natural pearls can form in both saltwater and freshwater, most are baroque shape.

Cultured pearls. Pearls produced in both fresh- and saltwater. A saltwater pearl is produced when a nucleus, a mother-of-pearl bead, and a section of mantle tissue from a mollusk, is introduced into a living mollusk. The mollusk views the nucleus as an irritant and coats it with concentric layers of nacre.

Blister pearl. This pearl is the result of an irritant lodged between the mantle tissue and the inner surface of the shell. The mollusk secretes nacre over the irritant just as it would if the irritant were in the mantle tissue. Blister pearls, often irregularly shaped, are cut from the shell and, therefore, are more or less flat. Blister pearls are not considered true pearls and the term "blister pearl" means a pearl occurring without outside interference by man.

Freshwater pearls. Pearls from mollusks living in freshwater lakes, rivers, and streams. Natural freshwater pearls are baroque in shape.

Freshwater cultured pearls. These are grown with the aid of man in lakes and streams. The two major culturing areas are Japan and China. Freshwater cultured pearls do not use a mother-of-pearl bead, but the nucleus is formed from mantle tissue alone. Lake Biwa, Japan is the most well known site for this culturing process. Freshwater cultured pearls are frequently dyed. The dye usually imparts much more vivid colors to the pearls than can be produced naturally. Most dyes are permanent. Freshwater pearls are now being cultured in the U.S. in the Tennessee river and some of its tributaries.

Keishi pearls. These are round natural freshwater pearls that weigh less than 1/4 of a pearl grain. A pearl grain is equal to 1/4 carat.

Mabe pearl. A half pearl cut off the shell. A mother-of-pearl bead is cemented to the inner shell of a mollusk and the mollusk covers it with nacre. When harvested the pearl is cut from the shell exposing the bead. The entire original bead is removed and the empty shell filled with a new bead cemented into place. Sometimes only a thin mother-of-pearl backing bead is used leaving most of the pearl hollow and as fragile as an empty egg shell.

BIBLIOGRAPHY

Aldred, Cyril. 1978. *Jewels of the pharaohs.* London: Thames and Hudson.

Altobelli, Cos. 1986. *Handbook of jewelry and gemstone appraising.* Los Angeles: American Gem Society.

Babcock, Henry A. 1980. *Appraisal principles and procedures.* Washington, D.C.: American Society of Appraisers.

Baker, Lillian. 1978. *One hundred years of collectible jewelry (1850–1950).* Paducah, KY: Collector Books.

———. 1981. *Art Nouveau & Art Deco jewelry.* Paducah, KY: Collector Books.

Becker, Vivienne. 1987. 2nd. Ed. *Antique and twentieth century jewellery.* Great Britain: N.A.G. Press Ltd.

Bell, Jeanenne. 1985. *Answers to questions about old jewelry (1840–1950).* Florence, AL: Books Americana, Inc.

———. 1985. *The appraisal of antique and period jewelry.* Bloomington: Indiana University.

Berk, Merle. 1988. Always in style. *Lapidary Journal* 42, no. 5, (August): 41.

Birren, Faber. 1963. *Color: From ancient mysticism to modern science.* New Jersey: Citadel Press.

Bhushan, Jamila Brij. 1964. *Indian jewellery, ornaments and decorative designs.* Bombay: D.B. Taraporevala Sons & Co.

Boardman, John. 1975. *Intaglios and rings, Greek, Etruscan and Eastern.* London: Thames and Hudson.

Bradford, Ernle. 1967. *Four centuries of European jewellery.* Feltham, Middlesex: Spring Books, Hamlyn House.

Bury, Shirley. 1984. *Rings.* Owings Mills, MD: Stemmer House.

———. 1985. *Sentimental jewellery.* Owings Mills, MD: Stemmer House.

Clark, Grahame. 1986. *Symbols of excellence.* London: Cambridge University Press.

Cocks, Anna Somers. 1980. *An introduction to courtly jewellery.* London: The Compton Press, Ltd.

Cohen, Neil, Joe Tenhagen, Donald Palmieri. 1983. *Master gemologist appraiser program.* Washington, DC: American Society of Appraisers.

Dubin, Lois Sherr. 1987. *The history of beads.* New York: Harry N. Abrams, Inc.

Dudley, Dorothy H., and Irma Bezold Wilkinson. 1979. *Museum registration methods.* Washington, DC: American Association of Museums.

Edwards, Charles. 1804. *History and poetry of finger-rings.* New York: John W. Lovell Co.

Egger, Gerhart. 1988. *Generations of jewelry.* West Chester, PA: Schiffer Publishing Ltd.

Evans, Joan. 1970. *A history of jewellery 1100–1870.* Boston, MA: Boston Book and Art.

Farn, Alexander. 1986. *Pearls: Natural, cultured and imitation.* London: Butterworths.

Falkiner, Richard. 1968. *Investing in antique jewelry.* New York: Clarkson N. Potter, Inc.

Flower, Margaret. 1951. *Victorian jewellery.* New York: Duell, Sloan and Pearce.

Garside, Anne, Ed. 1980. *Jewelry—ancient to modern.* New York: The Viking Press.

Gernsheim, Alison. 1981. *Victorian and Edwardian fashion.* Canada: General Publishing Co.

Harrison, Henry S., and Julie S. 1983. *Harrisons' illustrated dictionary of real estate appraisal.* New Haven, CT: Collegiate Distributing Co.

Hughes, Graham. 1972. *The art of jewelry.* New York: Viking Press, Inc.

Hyslop, K. R. Maxwell. 1971. *Western Asiatic jewellery, c.3000-612 B.C.* London: Methuen & Co. Ltd.

Ivory: an international history and illustrated survey. 1987. New York: Harry N. Abrams, Inc.

Jensen, Shelle S. 1988. Man and his jewelry history: Jewelry studs. *Modern Jeweler* 87, no. 8, (August): 64.

Jessup, Ronald. 1951. *Anglo-Saxon jewellery.* London: Faber and Faber.

Jokelson, Paul. 1968. *Sulphides, the art of cameo incrustation.* New York: Nelson & Sons.

Jolliff, Jim. 1988. *Gems and jewelry appraisal theory.* Annapolis, MD/Washington, DC: National Association of Jewelry Appraisers Conference.

Jonas, Joyce. 1988. The elegant Edwardians. *Heritage Jewelers' Circular Keystone* 154, no. 8, (August): 221–225.

Jones, William. 1968. *Finger-ring lore.* Detroit, MI: Singing Tree Press.

Kaplan, Arthur Guy. 1985. *The official price guide to antique jewelry.* Orlando, FL: House of Collectibles.

Kovel, Ralph and Terry. 1983. *The Kovels' antique & collectibles price list.* New York: Crown Publishers.

Kunz, George Frederick. 1917. *Rings for the finger.* Philadelphia, PA: J. B. Lippincott Co.

Lenzen, Godehard. 1970. *The history of diamond production and the diamond trade.* London: Barrie and Jenkins.

Letson, Neil. 1988. Art Deco jewelry—its past, present and future. *Jewelers' Circular Keystone* 154, no. 8, (August): 233–236.

Liddicoat, Richard T., and Lawrence L. Copeland, Eds. 1970. *The jeweler's manual.* Los Angeles, CA: Gemological Institute of America.

Liddicoat, Richard T. 1975. *Handbook of gem identification.* Los Angeles, California: Gemological Institute of America.

MacBride, Dexter D. 1976. *Commentary on personal property appraisal.* Appraisal Monograph #7. Washington, DC: American Society of Appraisers.

Mack, John. Ed. 1988. *Ethnic jewelry.* New York: Harry N. Abrams, Inc.

Marcum, David. 1986. *Dow Jones-Irwin guide to fine gems and jewelry.* Homewood, IL: Dow Jones-Irwin.

Martin, Deborah Dupont. 1987. Gemstone durability: Design to display. *Gems & Gemology,* 23, no. 2. (Summer): 63–77.

Mastai, Marie-Louise d'Otrange. 1981. *Jewelry.* Washington, DC: Smithsonian Institution.

Metropolitan Museum of Art. 1928. *Jewelry: The art of the goldsmith in classical times.* New York.

________. 1976. *Treasures of Tutankhamun.* New York: Ballantine Books.

________. 1980. *The imperial style: Fashions of the Hapsburg era.* New York.

Meyer, Franz S. 1987. *Meyer's handbook of ornament.* London: Omega Books.

Miller, Anna M. 1988. *Gems and jewelry appraising: Techniques of professional practice.* New York: Van Nostrand Reinhold.

Misiorowski, Elise B., and Dona M. Dirlam. 1986. Art nouveau: Jewels and jewelers. *Gems & Gemology,* 22, no. 4, (Winter): 209–228.

Muller, Pricilla E. 1972. *Jewels in Spain, 1500–1800.* New York: The Hispanic Society of America.

Mulvagh, Jane. 1988. *Costume jewelry in vogue.* New York: Thames and Hudson.

Munn, Geoffrey C. 1983. *Castellani and Giuliano.* New York: Rizzoli International Publications, Inc.

McCarthy, James Remington. 1945. *Rings through the ages.* New York: Harper & Brothers.

Ogden, Jack. 1982. *Jewellery of the ancient world.* New York: Rizzoli International Publications, Inc.

Ouspensky, Leonid, and Vladimir Lossky. 1983. *The meaning of icons.* New York: St. Vladimir's Seminary Press.

Oved, Sah. 1953. *The book of necklaces.* London: Arthur Barker.

Proddow, Penny, and Debra Healy. 1987. *American jewelry, glamour and tradition.* New York: Rizzoli International Publications, Inc.

Rickert, Richard. 1986. *Valuation: An interdisciplinary approach to appraisal.* Washington, D.C: American Society of Appraisers.

Rose, Augustus F., and Antonio Cirino. 1967. *Jewelry making and design.* New York: Dover Publications.

Ross, Heather Colyer. 1978. *Bedouin jewellery in Saudi Arabia.* London: Stacey International.

Sataloff, Joseph. 1984. *Art nouveau jewelry.* Bryn Mawr, PA: Dorrance & Co., Inc.

Scarisbrick, Diana. 1984. *The costume accessories series: Jewellery.* London: B.T. Batsford, Ltd.

Shirai, Shohei. 1981. *Pearls.* Japan: Marine Planning.

Stein, Nachum. 1981. *Evaluating diamonds: Beauty, value, investment.* New York: Hasenfeld-Stein, Inc.

Steingraber, Erich. 1957. *Antique jewelry.* New York: Frederick A. Praeger.

Tait, Hugh. Ed. 1986. *Jewelry 7,000 years.* New York: Harry N. Abrams.

Tenhagen, Joseph W. 1988. Appraising colored stones. *Lapidary Journal,* 41, no. 11 (February): 43–48.

Tenhagen, Joseph W. 1988. Contemporary jewelry. *Lapidary Journal,* 42, no. 5 (August): 49–52.

Tolansky, S. 1962. *The history and use of diamonds.* London: Methuen & Co., Ltd.

Vever, Henri. 1906. *La bijouterie Francaise au XIX siecle (1800–1900),* Vol. 1. Paris: H. Floury.

Villegas, Ramon N. 1983. *Kayamanan, the Philippine jewelry tradition.* Manila: Central Bank of the Philippines.

Ward, A., J. Cherry, C. Gere, B. Cartlidge. 1981. *Rings through the ages.* New York: Rizzoli International Publications, Inc.

Weber, Edith. 1988. Mysteries of antique jewelry. *Collectors Clocks & Jewelry,* 1, no. 1. (Fall): 46–49.

Wicks, Sylvia. 1985. *Jewelry making manual.* London: Quill Publishing Ltd.

Williams, C.A.S., 1976. *Outlines of Chinese symbolism & art motives.* New York: Dover Publications.

Zeitner, June Culp. 1988. The fine art of gem engraving. *Lapidary Journal,* 42, no. 8 (November): 22–32.

Zucker, Benjamin. 1976. *How to invest in gems: Everyone's guide to buying rubies, sapphires, emeralds, and diamonds.* New York: Quadrangle/The New York Times Book Co.

INDEX